ASTRONOMY

LECTOR HOUSE PUBLIC DOMAIN WORKS

ASTRONOMY

DAVID PECK TODD

ISBN: 978-93-5420-004-5

Published: 1922

LECTOR HOUSE LLP
E-MAIL: lectorpublishing@gmail.com

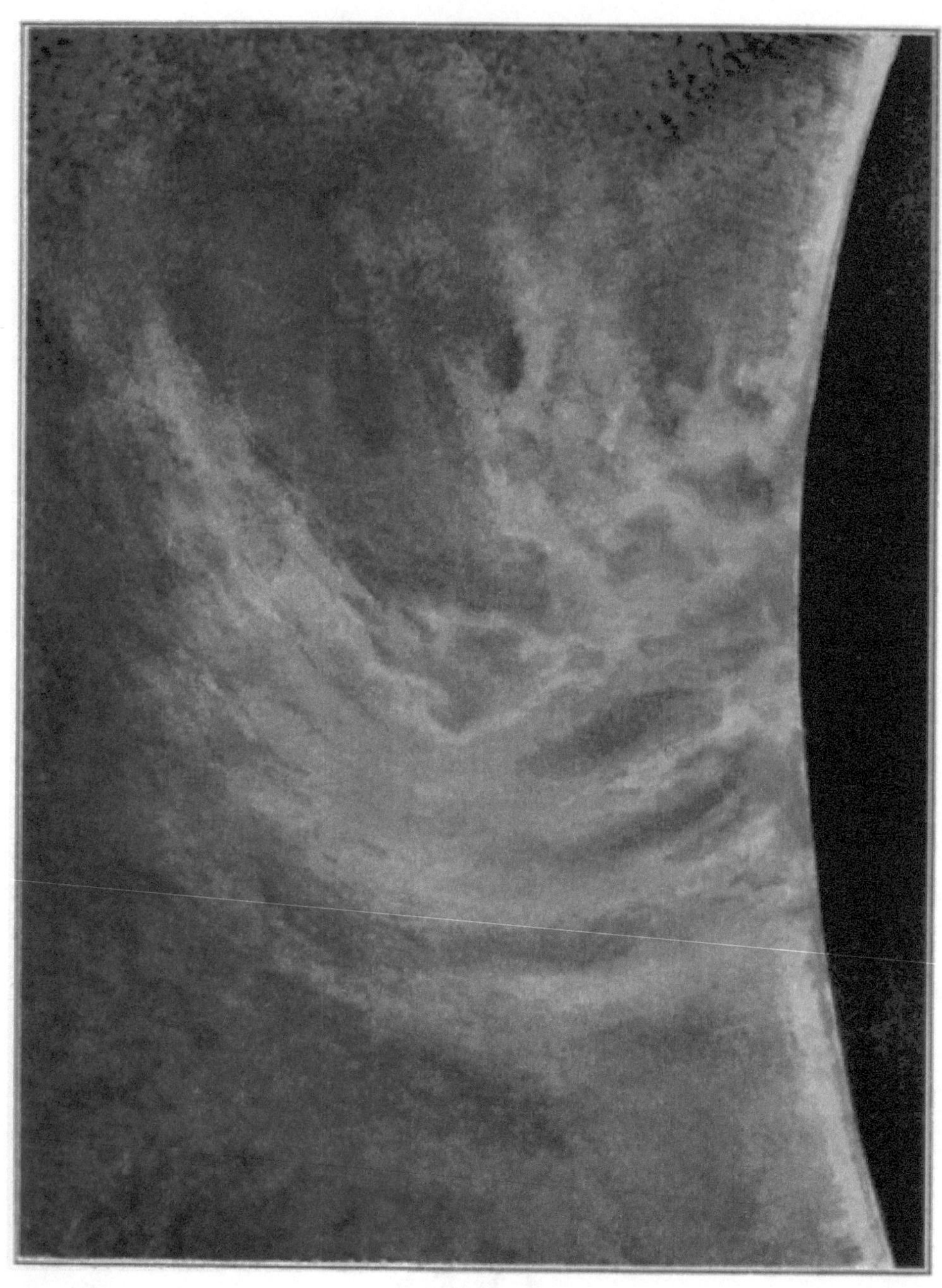

Photo, Mt. Wilson Solar Observatory

An Active Prominence of the Sun, 140,000 Miles High, photographed July 9, 1917.

ASTRONOMY
THE SCIENCE OF THE HEAVENLY BODIES

BY

DAVID TODD

Director Emeritus, Amherst College Observatory

MCMXXII

PREFACE

Sir William Rowan Hamilton, the eminent mathematician of Dublin, has, of all writers ancient and modern, most fittingly characterized the ideal science of astronomy as man's golden chain connecting the heavens to the earth, by which we "learn the language and interpret the oracles of the universe."

The oldest of the sciences, astronomy is also the broadest in its relations to human knowledge and the interests of mankind. Many are the cognate sciences upon which the noble structure of astronomy has been erected: foremost of all, geometry and the higher mathematics, which tell us of motions, magnitudes and distances; physics and chemistry, of the origin, nature, and destinies of planets, sun, and star; meteorology, of the circulation of their atmospheres; geology, of the structure of the moon's surface; mineralogy, of the constitution of meteorites; while, if we attack, even elementally, the fascinating, though perhaps forever unsolvable, problem of life in other worlds, the astronomer must invoke all the resources that his fellow biologists and their many-sided science can afford him.

The progress of astronomy from age to age has been far from uniform—rather by leaps and bounds: from the earliest epoch when man's planet earth was the center about which the stupendous cosmos wheeled, for whom it was created, and for whose edification it was maintained—down to the modern age whose discoveries have ascertained that even our stellar universe, the vast region of the solar domain, is but one of the thousands of island universes that tenant the inconceivable immensities of space.

Such results have been attainable only through the successful construction and operation of monster telescopes that bring to the eye and visualize on photographic plates the faintest of celestial objects which were the despair of astronomers only a few years ago.

But the end is not yet; astronomy to-day is but passing from infancy to youth. And with new and greater telescopes, with new photographic processes of higher sensitivity, with the help of modern invention in overcoming the obstacle of the air—that constant foe of the astronomer—who will presume to set down any limit to the leaps and bounds of astronomy in the future?

So rapid, indeed, has been the progress of astronomy in very recent years that the present is especially favorable for setting forth its salient features; and this book is an attempt to present the wide range of astronomy in readable fashion, as if a story with a definite plot, from its origin with the shepherds of ancient Chaldea down to present-day ascertainment of the actual scale of the universe, and definite measures of the huge volume of supersolar giants among the stars.

David Todd

Amherst College Observatory
November, 1921

CONTENTS

LIST OF ILLUSTRATIONS

CHAPTER I

ASTRONOMY A LIVING SCIENCE

Like life itself we do not know when astronomy began; we cannot conceive a time when it was not. Man of the early stone age must have begun to observe sun, moon, and stars, because all the bodies of the cosmos were there, then as now. With his intellectual birth astronomy was born.

Onward through the childhood of the race he began to think on the things he observed, to make crude records of times and seasons; the Chaldeans and Chinese began each their own system of astronomy, the causes of things and the reasons underlying phenomena began to attract attention, and astronomy was cultivated not for its own sake, but because of its practical utility in supplying the data necessary to accurate astrological prediction. Belief in astrology was universal.

The earth set in the midst of the wonders of the sky was the reason for it all. Clearly the earth was created for humanity; so, too, the heavens were created for the edification of the race. All was subservient to man; naturally all was geocentric, or earth-centered. From the savage who could count only to five, the digits of one hand, civilized man very slowly began to evolve; he noted the progress of the seasons; the old records of eclipses showed Thales, an early Greek, how to predict their happenings, and true science had its birth when man acquired the power to make forecasts that always came true.

Few ancient philosophers were greater than Pythagoras, and his conceptions of the order of the heavens and the shape and motion of the earth were so near the truth that we sometimes wonder how they could have been rejected for twenty centuries. We must remember, however, that man had not yet learned the art of measuring things, and the world could not be brought into subjection to him until he had. To measure he must have tools—instruments; to have instruments he must learn the art of working in metals, and all this took time; it was a slow and in large part imperceptible process; it is not yet finished.

The earliest really sturdy manifestation of astronomical life came with the birth of Greek science, culminating with Aristarchus, Hipparchus and Ptolemy. The last of these great philosophers, realizing that only the art of writing prevents man's knowledge from perishing with him, set down all the astronomical knowledge of that day in one of the three greatest books on astronomy ever written, the Almagest, a name for it derived through the Arabic, and really meaning "the greatest."

The system of earth and heaven seemed as if finished, and the authority of

Ptolemy and his Almagest were as Holy Writ for the unfortunate centuries that followed him. With fatal persistence the fundamental error of his system delayed the evolutionary life of the science through all that period.

But man had begun to measure. Geometry had been born and Eratosthenes had indeed measured the size of the earth. Tools in bronze and iron were fashioned closely after the models of tools of stone; astrolabes and armillary spheres were first built on geometric spheres and circles; and science was then laid away for the slumber of the Dark Ages.

Nevertheless, through all this dreary period the life of the youthful astronomical giant was maintained. Time went on, the heavens revolved; sun, moon, and stars kept their appointed places, and Arab and Moor and the savage monarchs of the East were there to observe and record, even if the world-mind was lying fallow, and no genius had been born to inspire anew that direction of human intellect on which the later growth of science and civilization depends. With the growth of the collective mind of mankind, from generation to generation, we note that ordered sequence of events which characterizes the development of astronomy from earliest peoples down to the age of Newton, Herschel, and the present. It is the unfolding of a story as if with a definite plot from the beginning.

Leaving to philosophical writers the great fundamental reason underlying the intellectual lethargy of the Dark Ages, we only note that astronomy and its development suffered with every other department of human activity that concerned the intellectual progress of the race. To knowledge of every sort the medieval spirit was hostile. But with the founding and growth of universities, a new era began. The time was ripe for Copernicus and a new system of the heavens. The discovery of the New World and the revival of learning through the universities added that stimulus and inspiration which marked the transition from the Middle Ages to our modern era, and the life of astronomy, long dormant, was quickened to an extraordinary development.

It fell to the lot of Copernicus to write the second great book on astronomy, "De Revolutionibus Orbium Cœlestium." But the new heliocentric or sun-centered system of Copernicus, while it was the true system bidding fair to replace the false, could not be firmly established except on the basis of accurate observation.

How fortunate was the occurrence of the new star of 1572, that turned the keen intellect of Tycho Brahe toward the heavens! Without the observational labors of Tycho's lifetime, what would the mathematical genius of Kepler have availed in discovery of his laws of motion of the planets?

Historians dwell on the destruction and violent conflicts of certain centuries of the Middle Ages, quite overlooking the constructive work in progress through the entire era. Much of this was of a nature absolutely essential to the new life that was to manifest itself in astronomy. The Arabs had made important improvements in mathematical processes, European artisans had made great advances in the manufacture of glass and in the tools for working in metals.

Then came Galileo with his telescope revealing anew the universe to mankind. It was the north of Italy where the Renaissance was most potent, recalling the vigorous life of ancient Greece. Copernicus had studied here; it was the home of Galileo. Columbus was a Genoese, and the compass which guided him to the Western World was a product of deft Italian artisans whose skill with that of their successors was now available to construct the instruments necessary for further progress in the accurate science of astronomical observation. Even before Copernicus, Johann Müller, better known as Regiomontanus, had imbibed the learning of the Greeks while studying in Italy, and founded an observatory and issued nautical almanacs from Nuremberg, the basis of those by which Columbus was guided over untraversed seas.

About this time, too, the art of printing was invented, and the interrelation of all the movements then in progress led up to a general awakening of the mind of man, and eventually an outburst in science and learning, which has continued to the present day. Naturally it put new life into astronomy, and led directly up from Galileo and his experimental philosophy to Newton and the *Principia*, the third in the trinity of great astronomical books of all time.

To get to the bottom of things, one must study intimately the history of the intellectual development of Europe through the fifteenth and sixteenth centuries. Many of the western countries were ruled by sovereigns of extraordinary vigor and force of character, and their activities tended strongly toward that firm basis on which the foundations of modern civilization were securely laid.

Contemporaneously with this era, and following on through the seventeenth century, came the measurements of the earth by French geodesists, the construction of greater and greater telescopes and the wonderful discoveries with them by Huygens, Cassini, and many others.

Most important of all was the application of telescopes to the instruments with which angles are measured. Then for the first time man had begun to find out that by accurate measures of the heavenly bodies, their places among the stars, their sizes and distances, he could attain to complete knowledge of them and so conquer the universe.

But he soon realized the insufficiency of the mathematical tools with which he worked—how unsuited they were to the solution of the problem of three bodies (sun, earth, and moon) under the Newtonian law of gravitation, let alone the problem of n-bodies, mutually attracting each the other; and every one perturbing the motion of every other one. So the invention of new mathematical tools was prosecuted by Newton and his rival Leibnitz, who, by the way, showed himself as great a man as mathematician: "taking mathematics," wrote Leibnitz, "from the beginning of the world to the times when Newton lived, what he had done was much the better half." Newton was the greatest of astronomers who, since the revival of learning, had observed the motions of the heavenly bodies and sought to find out why they moved.

Copernicus, Tycho Brahe, Galileo, Kepler, Newton, all are bound together as in a plot. Not one of them can be dissociated from the greatest of all discoveries. But Newton, the greatest of them all, revealed his greatness even more by saying: "If I have seen further than other men, it is because I have been standing on the shoulders of giants." Elsewhere he says: "All this was in the two plague years of 1665 and 1666 [he was then but twenty-four], for in those days I was in the prime of my age for invention, and minded mathematics and philosophy more than at any time since." All school children know these as the years of the plague and the fire; but very few, in school or out, connect these years with two other far-reaching events in the world's history, the invention of the infinitesimal calculus and the discovery of the law of gravitation.

We have passed over the name of Descartes, almost contemporary with Galileo, the founder of modern dynamics, but his initiation of one of the greatest improvements of mathematical method cannot be overlooked. This era was the beginning of the Golden Age of Mathematics that embraced the lives of the versatile Euler, equally at home in dynamics and optics and the lunar theory; of La Grange, author of the elegant "Mécanique Analytique"; and La Place, of the unparalleled "Mécanique Céleste." With them and a fully elaborated calculus Newton's universal law had been extended to all the motions of the cosmos. Even the tides and precession of the equinoxes and Bradley's nutation were accounted for and explained. Mathematical or gravitational astronomy had attained its pinnacle—it seemed to be a finished science: all who were to come after must be but followers.

The culmination of one great period, however, proved to be but the inception of another epoch in the development of the living science.

The greatest observer of all time, with a telescope built by his own hands, had discovered a great planet far beyond the then confines of the solar system. Mathematicians would take care of Uranus, and Herschel was left free to build bigger telescopes still, and study the construction of the stellar universe. Down to his day astronomy had dealt almost wholly with the positions and motions of the celestial bodies—astronomy was a science of *where*. To inquire *what* the heavenly bodies *are*, seemed to Herschel worthy of his keenest attention also. While "a knowledge of the construction of the heavens has always been the ultimate object of my observations," as he said, and his ingenious method of star-gauging was the first practicable attempt to investigate the construction of the sidereal universe, he nevertheless devoted much time to the description of nebulæ and their nature, as well as their distribution in space. He was the founder of double-star astronomy, and his researches on the light of the stars by the simple method of sequences were the inception of the vast fields of stellar photometry and variable stars. The physics of the sun, also, was by no means neglected; and his lifework earned for him the title of father of descriptive astronomy.

While progress and discovery in the earlier fields of astronomy were going on, the initial discoveries in the vast group of small planets were made at the beginning of the nineteenth century. The great Bessel added new life to the science by revolutionizing the methods and instruments of accurate observation, his work culminating in the measure of the distance of 61 Cygni, first of all the stars whose distance from the sun became known.

Wonderful as was this achievement, however, a greater marvel still was announced just before the middle of the century—a new planet far beyond Uranus, whose discovery was made as a direct result of mathematical researches by Adams and Le Verrier, and affording an extraordinary verification of the great Newtonian law. These were the days of great discoveries, and about this time the giant of all the astronomical tools of the century was erected by Lord Rosse, the "Leviathan" reflector with a speculum six feet in diameter, which remained for more than half a century the greatest telescope in the world, and whose epochal discovery of spiral nebulæ has greater significance than we yet know or perhaps even surmise.

The living science was now at the height of a vigorous development, when a revolutionary discovery was announced by Kirchhoff which had been hanging fire nearly half a century—the half century, too, which had witnessed the invention of photography, the steam engine, the railroad, and the telegraph: three simple laws by which the dark absorption lines of a spectrum are interpreted, and the physical and chemical constitution of sun and stars ascertained, no matter what their distance from us.

Huggins in England and Secchi in Italy were quick to apply the discovery to the stars, and Draper and Pickering by masterly organization have photographed and classified the spectra of many hundred thousand stars of both hemispheres, a research of the highest importance which has proved of unique service in studies of stellar movements and the structure of the universe by Eddington and Shapley, Campbell and Kapteyn, with many others who are still engaged in pushing our knowledge far beyond the former confines of the universe.

Few are the branches of astronomy that have not been modified by photography and the spectroscope. It has become a measuring tool of the first order of accuracy; measuring the speed of stars and nebulæ toward and from us; measuring the rotational speed of sun and planets, corona and Saturnian ring; measuring the distances of whole classes of stars from the solar system; measuring afresh even the distance of the sun—the yardstick of our immediate universe; measuring the drift of the sun with his entire family of planets twelve miles every second in the direction of Alpha Lyræ; and discovering and measuring the speed of binary suns too close together for our telescopes, and so making real the astronomy of the invisible.

Impatient of the handicap of a turbulent atmosphere, the living science has sought out mountain tops and there erected telescopes vastly greater than the "Leviathan" of a past century. There the sun in every detail of disk and spectrum is photographed by day, and stars with their spectra and the nebulæ by night. Great streams of stars are discovered and the speed and direction of their drift ascertained. The marvels of the spiral nebulæ are unfolded, their multitudinous forms portrayed and deciphered.

And their distances? And the distances of the still more wonderful clusters? Far, inconceivably far beyond the Milky Way. And are they "island universes"? And can man, the measurer, measure the distance of the "mainland" beyond?

CHAPTER II

THE FIRST ASTRONOMERS

Who were the first astronomers? And who wrote the first treatise on astronomy, oldest of the sciences?

Questions not easy to answer in our day. With the progress of archæological research, or inquiry into the civilization and monuments of early peoples, it becomes certain that man has lived on this planet earth for tens of thousands of years in the past as an intelligent, observing, intellectual being; and it is impossible to assign any time so remote that he did not observe and philosophize upon the firmament above.

We can hardly imagine a people so primitive that they would fail to regard the sun as "Lord of the Day," and therefore all important in the scheme of things terrestrial. Says Anne Bradstreet of the sun in her "Contemplations":

> What glory's like to thee?
> Soul of this world, this universe's eye,
> No wonder some made thee deity.

To the Babylonians belongs the credit of the oldest known work on astronomy. It was written nearly six thousand years ago, about B. C. 3800, by their monarch Sargon the First, King of Agade. Only the merest fragments of this historic treatise have survived, and they indicate the reverence of the Babylonians for the sun. Another work by Sargon is entitled "Omens," which shows the intimate relationship of astronomy to mysticism and superstitious worship at this early date, and which persists even at the present day.

As remotely as B. C. 3000, the sun-god Shamash and his wife Aya are carved upon the historic cylinders of hematite and lapis lazuli, and one of the oldest designs on these cylinders represents the sun-god coming out of the Door of Sunrise, while a porter is opening the Gate of the East. The Semitic religion had as its basis a reverence for the bodies of the sky; and Samson, Hebrew for sun, was probably the sun-god of the Hebrews. The Phœnician deity, Baal, was a sun-god under differing designations; and at the epoch of the Shepherd Kings, about B. C. 1500, during the Hyksos dynasty, the sun-god was represented by a circle or disk with extended rays ending in hands, possibly the precursor of the frequently recurring Egyptian design of the winged disk or winged solar globe. Hittites, Persians, and Assyrians, as well as the Phœnicians, frequently represented the sun-god in similar fashion in their sacred glyphs or carvings.

For a long period in early human history, astronomy and astrology were pretty much the same. We can trace the history of astrology back as far as B. C. 3000 in ancient Babylonia. The motions of the sun, moon, and the five lucid planets of that time indicated the activity of the various gods who influenced human affairs. So the Babylonian priests devised an elaborate system of interpreting the phenomena of the heavens; and attaching the proper significance in human terms to everything that took place in the sky. In Babylonia and Assyria it was the king and his people for whom the prognostications were made out. It was the same in Egypt. Later, about the fifth century B. C., astrology spread through Greece, where astrologers developed the idea of the influence of planets upon individual concerns. Astrology persisted through the Dark Ages, and the great astronomers Copernicus, Tycho, Kepler, Gassendi, and Huygens were all astrologers as well. Milton makes many references to planetary influence, our language has many words with a direct origin in astrology, and in our great cities to-day are many astrologers who prepare individual horoscopes of more than ordinary interest.

It is difficult to assign the antiquity of the Chinese astronomy with any approach to definiteness. Their earliest records appear to have been total eclipses of the sun, going back nearly 2,200 years before the Christian era; and nearly a thousand years earlier the Hindu astronomy sets down a conjunction of all the planets, concerning which, however, there is doubt whether it was actually observed or merely calculated backward. Owing to a colossal misfortune, the burning of all native scientific books by order of the Emperor Tsin-Chi-Hwang-Ti, in B. C. 221, excepting only the volumes relating to agriculture, medicine, and astrology, the Chinese lost a precious mass of astronomical learning, accumulated through the ages. No less an authority than Wells Williams credits them with observing 600 solar eclipses between B. C. 2159 and A. D. 1223, and there must have been some centuries of eclipses observed and recorded anterior to B. C. 2159, as this is the date assigned to the eclipse which came unheralded by the astronomers royal, Hi and Ho, who had become intoxicated and forgot to warn the Court, in accord with their duty. China was thereby exposed to the anger of the gods, and Hi and Ho were executed by his Majesty's command. It is doubtful if there is an earlier record of any celestial phenomenon.

CHAPTER III

PYRAMID, TOMB, AND TEMPLE

Inquiry into the beginnings of astronomy in ancient Egypt reveals most interesting relations of the origins of the science to the life and work and worship of the people. Their astronomers were called the "mystery teachers of heaven"; their monuments indicate a civilization more or less advanced; and their temples were built on astronomical principles and dedicated to purpose of worship. The Egyptian records carry us back many thousands of years, and we find that in Egypt, as in other early civilizations, observation of the heavenly bodies may be embraced in three pretty distinct stages. Awe, fear, wonder and worship were the first. Then came utility: a calendar was necessary to tell men when "to plow and sow, to reap and mow," and a calendar necessitated astronomical observations of some sort. Following this, the third direction required observations of celestial positions and phenomena also, because astrology, in which the potentates of every ancient realm believed, could only thrive as it was based on astronomy.

Sun worship was preeminent in early Egypt as in India, where the primal antithesis between night and day struck terror in the unformed mind of man. In one of the Vedas occurs this significant song to the god of day: "Will the Sun rise again? Will our old friend the Dawn come back again? Will the power of Darkness be conquered by the God of Light?"

Quite different from India, however, is Egypt in matters of record: in India, records in papyrus, but no monuments of very great antiquity; in Egypt, no papyrus, but monuments of exceeding antiquity in abundance. Herodotus and Pliny have told us of the great antiquity of these monuments, even in their own day, and research by archæologist and astronomer has made it certain that the pyramids were built by a race possessing great knowledge of astronomy. Their temples, too, were constructed in strict relation to stars. Not only are the temples, as Edfu and Denderah, of exceeding interest in themselves, but associated with them are often huge monoliths of syenite, obelisks of many hundred tons in weight, which the astronomer recognizes as having served as observation pillars or gnomons. Specimens of these have wandered as far from home as Central Park and the bank of the Thames. But there is an even more remarkable wealth of temple inscriptions, zodiacs especially.

Next to the sun himself was the worship of the Dawn and Sunrise, the great revelations of nature. There were numerous hymns to the still more numerous sun-gods and the powers of sunlight. Ra was the sun-god in his noontide strength;

Osiris, the dying sun of sunset. Only two gods were associated with the moon, and for the stars a special goddess, Sesheta. Sacrifices were made at day-break; and the stars that heralded the dawn were the subjects of careful observation by the sacrificial priests, who must therefore have possessed a good knowledge of star places and names, doubtless in belts of stars extending clear around the heavens. These decans, as they were called, are the exact counterparts of the moon stations devised by the Arabians, Indians, and other peoples for a like purpose.

The plane or circle of observation, both in Egypt and India, was always the horizon, whether the sun was observed or moon or stars. So the sun was often worshiped by the ancient Egyptians as the "Lord of the Two Horizons." It is sometimes difficult to keep in mind the fact, in regard to all temples of the ancients, whether in Egypt or elsewhere, that in studying them we must deal with the risings or settings of the heavenly bodies in quite different fashion from that of the astronomer of to-day, who is mainly concerned only with observing them on the meridian. The axis of the temple shows by its direction the place of rising or setting: if the temple faces directly east or west, its amplitude is 0. Now the sun, moon, and planets are, as everyone knows, very erratic as to their amplitudes (i. e., horizon points) of rising and setting; so it must have been the stars that engrossed the attention of the earliest builders of temples. After that, temples were directed to the rising sun, at the equinox or solstices. Then came the necessity of finding out about the inclination or obliquity of the ecliptic, and this is where the gnomon was employed.

At Karnak are many temples of the solstitial order: the wonderful temple of Amen-Ra is so oriented that its axis stands in amplitude 26 degrees north of west, which is the exact amplitude of the sun at Thebes at sunset of the summer solstice. The axis of a lesser temple adjacent points to 26 degrees south of east, which is the exact amplitude of sunrise at the winter solstice. At Gizeh we find the temples oriented, not solstitially, but by the equinoxes, that is, they face due east and west. Peoples who worshiped the sun at the solstice must have begun their year at the solstice; and Sir Norman Lockyer shows how the rise of the Nile, which took place at the summer solstice, dominated not only the industry but the astronomy and religion of Egypt.

Looking into the question of temple orientation in other countries, as China, for example, Lockyer finds that the most important temple of that country, the Temple of the Sun at Peking, is oriented to the winter solstice; and Stonehenge, as has long been known, is oriented to sunrise at the summer solstice.

In like fashion the rising and setting of many stars were utilized by the Egyptians, in both temple and pyramid; and no astronomer who has ever seen these ancient structures and studied their orientations can doubt that they were built by astronomers for use by astronomers of that day. The priests were the astronomers, and the temples had a deep religious significance, with a ceremony of exceeding magnificence wherever observations of heavenly bodies were undertaken, whether of sun or stars.

Hindu and Persian astronomy must be passed over very briefly. Interesting as their systems are historically, there were few, if any, original contributions of importance, and the Indian treatises bear strong evidence of Greek origin.

CHAPTER IV
ORIGIN OF GREEK ASTRONOMY

While the Greeks laid the foundations of modern scientific astronomy, they were not as a whole observers: rather philosophers, we should say. The later representatives of the Greek School, however, saw the necessity of observation as a basis of true induction; and they discovered that real progress was not possible unless their speculative ideas were sufficiently developed and made definite by the aid of geometry, so that they became capable of detailed comparison with observation. This was the necessary and ultimate test with them, and the same is true to-day. The early Greek philosophers were, however, mainly interested, not in observations, but in guessing the causes of phenomena.

Thales of Miletus, founder of the Ionian School, introduced the system of Egyptian astronomy into Greece, about the end of the seventh century B. C. He is universally known as the first astronomer who ever predicted a total eclipse of the sun that happened when he said it would: the eclipse of B. C. 585. This he did by means of the Chaldean eclipse cycle of 18 years known as the Saros.

Aristarchus of Samos was the first and most eminent of the Alexandrian astronomers, and his treatise "On the Magnitudes and Distances of the Sun and Moon" is still extant. This method of ascertaining how many times farther the sun is than the moon is very simple, and geometrically exact. Unfortunately it is impossible, even to-day, to observe with accuracy the precise time when the moon "quarters," (an observation essential to his method), because the moon's terminal, or line between day and night, is not a straight line as required by theory, but a jagged one. By his observation, the sun was only twenty times farther away than the moon, a distance which we know to be nearly twenty times too small.

His views regarding other astronomical questions were right, although they found little favor among contemporaries. Not only was the earth spherical, he said, but it rotated on its axis and also traveled round the sun. Aristarchus was, indeed, the true originator of the modern doctrine of motions in the solar system, and not Copernicus, seventeen centuries later; but Seleucus appears to have been his only follower in these very advanced conceptions. Aristarchus made out the apparent diameters of sun and moon as practically equal to one another, and inferred correctly that their real diameters are in proportion to their distances from the earth. Also he estimated, from observations during an eclipse of the moon, that the moon's diameter is about one-third that of the earth. Aristarchus appears to have been one of the clearest and most accurate thinkers among the ancient astronomers; even his views concerning the distances of the stars were in accord with the fact that they are immeasurably distant as compared with the distances of the sun, moon, and planets.

Practically contemporary with Aristarchus were Timocharis and Aristillus, who were excellent observers, and left records of position of sun and planets which were exceedingly useful to their successors, Hipparchus and Ptolemy in particular. Indeed their observations of star positions were such that, in a way, they deserve the fame of having made the first catalogue, rather than Hipparchus, to whom is universally accorded that honor.

Spherical astronomy had its origin with the Alexandrian school, many famous geometers, and in particular Euclid, pointing the way. Spherics, or the doctrine of the sphere, was the subject of numerous treatises, and the foundations were securely laid for that department of astronomical research which was absolutely essential to farther advance. The artisans of that day began to build rude mechanical adaptations of the geometric conceptions as concrete constructions in wood and metal, and it became the epoch of the origin of astrolabes and armillary spheres.

CHAPTER V

MEASURING THE EARTH—ERATOSTHENES

All told, the Greek philosophers were probably the keenest minds that ever inhabited the planet, and we cannot suppose them so stupid as to reject the doctrine of a spherical earth. In fact so certain were they that the earth's true figure is a sphere that Eratosthenes in the third century B. C. made the first measure of the dimensions of the terrestrial sphere by a method geometrically exact.

At Syene in Upper Egypt the sun at the summer solstice was known to pass through the zenith at noon, whereas at Alexandria Eratosthenes estimated its distance as seven degrees from the zenith at the same time. This difference being about one-fiftieth of the entire circumference of a meridian, Eratosthenes correctly inferred that the distance between Alexandria and Syene must be one-fiftieth of the earth's circumference. So he measured the distance between the two and found it 5,000 stadia. This figured out the size of the earth with a percentage of error surprisingly small when we consider the rough means with which Eratosthenes measured the sun's zenith distance and the distance between the two stations.

Greatest of all the Greek astronomers and one of the greatest in the history of the science was Hipparchus who had an observatory at Rhodes in the middle of the second century B. C. His activities covered every department of astronomy; he made extensive series of observations which he diligently compared with those handed down to him by the earlier astronomers, especially Aristillus and Timocharis. This enabled him to ascertain the motion of the equinoxial points, and his value of the constant of precession of the equinoxes is exceedingly accurate for a first determination.

In 134 B. C. a new star blazed out in the constellation Scorpio, and this set Hipparchus at work on a catalogue of the brighter stars of the firmament, a monumental work of true scientific conception, because it would enable the astronomers of future generations to ascertain what changes, if any, were taking place in the stellar universe. There were 1,080 stars in his catalogue, and he referred their positions to the ecliptic and the equinoxes. Also he originated the present system of stellar magnitudes or orders of brightness, and his catalogue was in use as a standard for many centuries.

Hipparchus was a great mathematician as well, and he devoted himself to the improvement of the method of applying numerical calculations to geometrical figures: trigonometry, both plane and spherical, that is; and by some authorities he is regarded as the inventor of original methods in trigonometry. The system of spheres of Eudoxus did not satisfy him, so he devised a method of representing the paths of the heavenly bodies by perfectly uniform motion in circles. There is slight evidence that Apollonius of Perga may have been the originator of the system, but it was reserved for Hipparchus to work it out in final form. This enabled him to ascertain the varying length of the seasons, and he fixed the true length of the year as 365¼ days. He had almost equal success in dealing with the irregularities of the moon's motion, although the problem is much more complicated. The distance and size of the moon, by the method of Aristarchus, were improved by him, and he worked out, for the distance of the sun, 1,200 radii of the earth—a classic for many centuries.

Hipparchus devoted much attention to eclipses of both sun and moon, and we owe to him the first elucidation of the subject of parallax, or the effect of difference of position of an observer on the earth's surface as affecting the apparent projection of the moon against the sun when a solar eclipse takes place; whereas an eclipse of the moon is unaffected by parallax and can be seen at the same time by observers everywhere, no matter what their location on the earth. Indeed, with all that Hipparchus achieved, we need not be surprised that astronomy was regarded as a finished science, and made practically no progress whatever for centuries after his time.

Then came Claudius Ptolemæus, generally known as Ptolemy, the last great name in Greek astronomy. He lived in Alexandria about the middle of the second century A. D. and wrote many minor astronomical and astrological treatises, also works on geography and optics, in the last of which the atmospheric refraction of rays of light from the heavenly bodies, apparently elevating them toward the zenith, is first dealt with in true form.

CHAPTER VI

PTOLEMY AND HIS GREAT BOOK

Ptolemy was an observer of the heavens, though not of the highest order; but he had all the work of his predecessors, best of all Hipparchus, to build upon. Ptolemy's greatest work was the "Megale Syntaxis," generally known as the Almagest. It forms a nearly complete compendium of the ancient astronomy, and although it embodies much error, because built on a wrong theory, the Almagest nevertheless is competent to follow the motions of all the bodies in the sky with a close approach to accuracy, even at the present day. This marvelous work written at this critical epoch became as authoritative as the philosophy of Aristotle, and for many centuries it was the last word in the science. The old astrology held full sway, and the Ptolemaic theory of the universe supplied everything necessary: further progress, indeed, was deemed impossible.

The Almagest comprises in all thirteen books, the first two of which deal with the simpler observations of the celestial sphere, its own motion and the apparent motions of sun, moon, and planets upon it. He discusses, too, the postulates of his system and exhibits great skill as an original geometer and mathematician. In the third book he takes up the length of the year, and in the fourth book similarly the moon and the length of the month. Here his mathematical powers are at their best, and he made a discovery of an inequality in the moon's motion known as the evection. Book five describes the construction and use of the astrolabe, a combination of graduated circles with which Ptolemy made most of his observations. In the sixth book he follows mainly Hipparchus in dealing with eclipses of sun and moon. In the seventh and eighth books he discusses the motion of the equinox, and embodies a catalogue of 1,028 stars, substantially as in Hipparchus. The five remaining books of the Almagest deal with the planetary motions, and are the most important of all of Ptolemy's original contributions to astronomy. Ptolemy's fundamental doctrines were that the heavens are spherical in form, all the heavenly motions being in circles. In his view, the earth too is spherical, and it is located at the center of the universe, being only a point, as it were, in comparison. All was founded on mere appearance combined with the philosophical notion that the circle being the only perfect curve, all motions of heavenly bodies must take place in earth-centered circles. For fourteen or fifteen centuries this false theory persisted, on the authority of Ptolemy and the Almagest, rendering progress toward the development of the true theory impossible.

Ptolemy correctly argued that the earth itself is a sphere that is curved from east to west, and from north to south as well, clinching his argument, as we do to-day, by the visibility of objects at sea, the lower portions of which are at first concealed from our view by the curved surface of the water which intervenes. To Ptolemy also the earth is at the center of the celestial sphere, and it has no motion of translation from that point; but his argument fails to prove this. Truth and error, indeed, are so deftly intermingled that one is led to wonder why the keen intelligence of this great philosopher permitted him to reject the simple doctrine of the earth's rotation on its axis. But if we reflect that there was then no science of natural philosophy or physics proper, and that the age was wholly undeveloped along the lines of practical mechanics, we shall see why the astronomers of Ptolemy's time and subsequent centuries were content to accept the doctrines of the heavens as formulated by him.

When it came to explaining the movements of the "wandering stars," or planets, as we term them, the Ptolemaic theory was very happy in so far as accuracy was concerned, but very unhappy when it had to account for the actual mechanics of the cosmos in space. Sun and moon were the only bodies that went steadily onward, easterly: whereas all the others, Mercury, Venus, Mars, Jupiter, Saturn, although they moved easterly most of the time, nevertheless would at intervals slow down to stationary points, where for a time they did not move at all, and then actually go backward to the west, or retrograde, then become stationary again, finally resuming their regular onward motion to the east.

To help out of this difficulty, the worst possible mechanical scheme was invented, that known as the epicycle. Each of the five planets was supposed to have a fictitious "double," which traveled eastward with uniformity, attached to the end of a huge but mechanically impossible bar. The earth-centered circle in which this traveled round was called the "deferent." What this bar was made of, what stresses it would be subjected to, or what its size would have to be in order to keep from breaking—none of these questions seems to have agitated the ancient and medieval astronomers, any more than the flat-earth astronomy of the Hindu is troubled by the necessity of something to hold up the tortoise that holds up the elephant that holds up the earth.

But at the end of this bar is jointed or swiveled another shorter bar, to the revolving end of which is attached the actual planet itself; and the second bar, by swinging once round the end of the primary advancing bar, would account for the backward or retrograde motion of the planet as seen in the sky. For every new irregularity that was found, in the motion of Mars, for instance, a new and additional bar was requisitioned, until interplanetary space was hopelessly filled with revolving bars, each producing one of the epicycles, some large, some small, that were needed to take up the vagaries of the several planets.

The Arabic astronomers who kept the science alive through the Middle Ages added epicycle to epicycle, until there was every justification for Milton's verses descriptive of the sphere:

> With Centric and Eccentric scribbled o'er,
> Cycle and Epicycle, Orb in Orb.

CHAPTER VII

ASTRONOMY OF THE MIDDLE AGES

With the fall of Alexandria and the victory of Mohammed throughout the West, and a consequent decline in learning, supremacy in science passed to the East and centered round the caliphs of Bagdad in the seventh and eighth centuries. They were interested in astronomy only as a practical, and to them useful, science, in adjusting the complicated lunar calendar of the Mohammedans, in ascertaining the true direction of Mecca which every Mohammedan must know, and in the revival of astrology, to which the Greeks had not attached any particular significance.

Harun al-Rashid ordered the Almagest and many other Greek works translated, of which the modern world would otherwise no doubt never have heard, as the Greek originals are not extant.

Splendid observatories were built at Damascus and Bagdad, and fine instruments patterned after Greek models were continuously used in observing. The Arab astronomers, although they had no clocks, were nevertheless so fully impressed with the importance of time that they added extreme value to their observations of eclipses, for example, by setting down the altitudes of sun or stars at the same time. On very important occasions the records were certified on oath by a body of barristers and astronomers conjointly—a precedent which fortunately has never been followed.

About the middle of the ninth century, the Caliph Al-Mamun directed his astronomers to revise the Greek measures of the earth's dimensions, and they had less reverence for the Almagest than existed in later centuries: indeed, Tabit ben Korra invented and applied to the tables of the Almagest a theoretical fluctuation in the position of the ecliptic which he called "trepidation," which brought sad confusion into astronomical tables for many succeeding centuries.

Albategnius was another Arab prince whose record in astronomy in the ninth and tenth centuries was perhaps the best: the Ptolemaic values of the precession of the equinoxes and of the obliquity of the ecliptic were improved by new observations, and his excellence as mathematician enabled him to make permanent improvements in the astronomical application of trigonometry.

Abul Wefa was the last of the Bagdad astronomers in the latter half of the tenth century, and his great treatise on astronomy known as the Almagest is sometimes confused with Ptolemy's work. Following him was Ibn Yunos of Cairo, whose labors culminated in the famous Hakemite Tables, which became the standard in mathematical and astronomical computations for several centuries.

Mohammedan astronomy thrived, too, in Spain and northern Africa. Arzachel of Toledo published the Toledan Tables, and his pupils made improvements in instruments and the methods of calculation. The Giralda was built by the Moors in Seville in 1196, the first astronomical observatory on the continent of Europe; but within the next half century both Seville and Cordova became Christian again, and Arab astronomy was at an end.

Through many centuries, however, the science had been kept alive, even if no great original advances had been achieved; and Arab activities have modified our language very materially, adding many such words as almanac, zenith, and radii, and a wealth of star names, as Aldebaran, Rigel, Betelgeuse, Vega, and so on.

Meanwhile, other schools of astronomy had developed in the East, one at Meraga near the modern Persia, where Nassir Eddin, the astronomer of Hulagu Khan, grandson of the Mongol emperor Genghis Khan, built and used large and carefully constructed instruments, translated all the Greek treatises on astronomy, and published a laborious work known as the Ilkhanic Tables, based on the Hakemite Tables of Ibn Yunos.

More important still was the Tartar school of astronomy under Ulugh Beg, a grandson of Tamerlane, who built an observatory at Samarcand in 1420, published new tables of the planets, and made with his excellent instruments the observations for a new catalogue of stars, the first since Hipparchus, the star places being recorded with great precision.

The European astronomy of the Middle Ages amounted to very little besides translation from the Arabic authors into Latin, with commentaries. Astronomers under the patronage of Alfonso X of Leon and Castile published in 1252 the Alfonsine Tables, which superseded the Toledan tables and were accepted everywhere throughout Europe. Alfonso published also the "Libros del Saber," perhaps the first of all astronomical cyclopedias, in which is said to occur the earliest diagram representing a planetary orbit as an ellipse: Mercury's supposed path round the earth as a center.

Purbach of Vienna about the middle of the 15th century began his "Epitome of Astronomy" based on the "Almagest" of Ptolemy, which was finished by his collaborator Regiomontanus, who was an expert in mathematics and published a treatise on trigonometry with the first table of sines calculated for every minute from $0°$ to $90°$, a most helpful contribution to theoretical astronomy.

Regiomontanus had a very picturesque career, finally taking up his residence in Nuremberg, where a wealthy citizen named Walther became his patron, pupil, and collaborator. The artisans of the city were set at work on astronomical instruments of the greatest accuracy, and the comet of 1472 was the first to be observed and studied in true scientific fashion. Regiomontanus was very progressive and the invention of the new art of printing gave him an opportunity to publish Purbach's treatise, which went through several editions and doubtless had much to do in promoting dissatisfaction with the ancient Ptolemaic system, and was thus most significant in preparing a background for the coming of the new Copernican order.

The Nuremberg presses popularized astronomy in other important ways, issuing almanacs, the first precursors of our astronomical Ephemerides. Regiomontanus was practical as well, and invented a new method of getting a ship's position at sea, with tables so accurate that they superseded all others in the great voyages of discovery, and it is probable that they were employed by Columbus in his discovery of the American continent. Regiomontanus had died several years earlier, in 1475 at Rome, where he had gone by invitation of the Pope to effect a reformation in the calendar. He was only forty, and his patron Walther kept on with excellent observations, the first probably to be corrected for the effect of atmospheric refraction, although its influence had been known since Ptolemy. The Nuremberg School lasted for nearly two centuries.

Nearly contemporary with Regiomontanus were Fracastoro and Peter Apian, whose original observations on comets are worthy of mention because they first noticed that the tails of these bodies always point away from the sun. Leonardo da Vinci was the first to give the true explanation of earth-shine on the moon, and similarly the moon-illumination of the earth; and this no doubt had great weight in disposing of the popular notion of an essential difference of nature between the earth and celestial bodies—all of which helped to prepare the way for Copernicus and the great revolution in astronomical thought.

CHAPTER VIII

COPERNICUS AND THE NEW ERA

Throughout the Middle Ages the progress of astronomy was held back by a combination of untoward circumstances. A prolonged reaction from the heights attained by the Greek philosophers was to be expected. The uprising of the Mohammedan world, and the savage conquerors in the East did not produce conditions favorable to the origin and development of great ideas.

At the birth of Copernicus, however, in 1473, the time was ripening for fundamental changes from the ancient system, the error of which had helped to hold back the development of the science for centuries. The fifteenth century was most fruitful in a general quickening of intelligence, the invention of printing had much to do with this, as it spread a knowledge of the Greek writers, and led to conflict of authorities. Even Aristotle and Ptolemy were not entirely in harmony, yet each was held inviolate. It was the age of the Reformation, too, and near the end of the century the discovery of America exerted a powerful stimulus in the advance of thought.

Copernicus searched the works of the ancient writers and philosophers, and embodied in this new order such of their ideas as commended themselves in the elaboration of his own system.

Pythagoras alone and his philosophy looked in the true direction. Many believe that he taught that the sun, not the earth, is at the center of our solar system; but his views were mingled with the speculative philosophy of the Greeks, and none of his writings, barring a few meager fragments, have come down to our modern age.

To many philosophers, through all these long centuries, the true theory of the celestial motions must have been obvious, but their views were not formulated, nor have they been preserved in writing. So the fact remains that Copernicus alone first proved the truth of the system which is recognized to-day. This he did in his great treatise entitled "De Revolutionibus Orbium Cœlestium," the first printed copy of which was dramatically delivered to him on his deathbed, in May, 1543. The seventy years of his life were largely devoted to the preparation of this work, which necessitated many observations as well as intricate calculations based upon them. Being a canon in the church, he naturally hesitated about publishing his revolutionary views, his friend Rheticus first doing this for him in outline in 1540.

So simple are the great principles that they may be embodied in very few words; what appears to us as the daily revolution of the heavens is not a real motion, but only an apparent one; that is, the heavens are at rest, while the earth itself is in motion, turning round an axis which passes through its center. And the second proposition is that the earth is simply one of the six known planets; and they all revolve round the sun as the true center. The solar system, therefore, is "heliocentric," or sun-centered, not "geocentric" or earth-centered, as taught by the Ptolemaic theory.

Copernicus demonstrates clearly how his system explains the retrograde motion of the planets and their stationary points, no matter whether they are within the orbit of the earth, as Mercury and Venus, or outside of it, as Mars, Jupiter, and Saturn. His system provides also the means of ascertaining with accuracy the proportions of the solar system, or the relative distances of the planets from the sun and from each other. In this respect also his system possessed a vast advantage over that of Ptolemy, and the planetary distances which Copernicus computed are very close approximations to the measures of the present day.

Reinhold revised the calculations of Copernicus and prepared the "Tabulæ Prutenicæ," based on the "De Revolutionibus," which proved far superior to the Alfonsine Tables, and were only supplanted by the Rudolphine Tables of Kepler. On the whole we may regard the lifework of Copernicus as fundamentally the most significant in the history and progress of astronomy.

CHAPTER IX

TYCHO, THE GREAT OBSERVER

Clear as Copernicus had made the demonstration of the truth of his new system, it nevertheless failed of immediate and universal acceptance. The Ptolemaic system was too strongly intrenched, and the motions of all the bodies in the sky were too well represented by it. Accurate observations were greatly needed, and the Landgrave William IV. of Hesse built the Cassel Observatory, which made a new catalogue of stars, and introduced the use of clocks to carry on the time as measured by the uniform motion of the celestial sphere. Three years after the death of Copernicus, Tycho Brahe was born, and when he was 30 the King of Denmark built for him the famous observatory of Uraniborg, where the great astronomer passed nearly a quarter of a century in critically observing the positions of the stars and planets. Tycho was celebrated as a designer and constructor of new types of astronomical instruments, and he printed a large volume of these designs, which form the basis of many in use at the present day. Unfortunately for the genius of Tycho and the significance of his work, the invention of the telescope had not yet been made, so that his observations had not the modern degree of accuracy. Nevertheless, they were destined to play a most important part in the progress of astronomy.

Tycho was sadly in error in his rejection of the Copernican system, although his reasons, in his day, seemed unanswerable. If the outer planets were displaced among the stars by the annual motion of the earth round the sun, he argued, then the fixed stars must be similarly displaced—unless indeed they be at such vast distances that their motions would be too slight to be visible. Of course we know now that this is really true, and that no instruments that Tycho was able to build could possibly have detected the motions, the effects of which we now recognize in the case of the nearer fixed stars in their annual, or parallactic, orbits.

The remarkably accurate instruments devised by Tycho Brahe and employed by him in improving the observations of the positions of the heavenly bodies were no doubt built after descriptions of astrolabes such as Hipparchus used, as described by Ptolemy. In his "Astronomiæ Instauratæ Mechanica" we find illustrations and descriptions of many of them.

One is a polar astrolabe, mounted somewhat as a modern equatorial telescope is, and the meridian circle is adjustable so that it can be used in any place, no matter what its latitude might be. There is a graduated equatorial ring at right angles to the polar axis, so that the astrolabe could be used for making observations outside the meridian as well as on it. This equatorial circle slides through grooves, and is furnished with movable sights, and a plumb line from the zenith or highest point of the meridian circle makes it possible to give the necessary adjustment in the vertical. Screws for adjustment at the bottom are provided, just as in our modern instruments, and two observers were necessary, taking their sights simultaneously; unless, as in one type of the instrument, a clock, or some sort of measure of time, was employed.

Another early type of instrument is called by Tycho the ecliptic astrolabe (*Armillæ Zodiacales*, or the Zodiacal Rings). It resembles the equatorial astrolabe somewhat, but has a second ring inclined to the equatorial one at an angle equal to the obliquity of the ecliptic. In observing, the equatorial ring was revolved round till the ecliptic ring came into coincidence with the plane of the ecliptic in the sky. Then the observation of a star's longitude and latitude, as referred to the ecliptic plane, could be made, quite as well as that of right ascension and declination on the equatorial plane. But it was necessary to work quickly, as the adjustment on the ecliptic would soon disappear and have to be renewed.

Tycho is often called the father of the science of astronomical observation, because of the improvements in design and construction of the instruments he used. His largest instrument was a mural quadrant, a quarter-circle of copper, turning parallel to the north-and-south face of a wall, its axis turning on a bearing fixed in the wall. The radius of this quadrant was nine feet, and it was graduated or divided so as to read the very small angle of ten seconds of arc—an extraordinary degree of precision for his day.

Tycho built also a very large alt-azimuth quadrant, of six feet radius. Its operation was very much as if his mural quadrant could be swung round in azimuth. At several of the great observatories of the present day, as Greenwich and Washington, there are instruments of a similar type, but much more accurate, because the mechanical work in brass and steel is executed by tools that are essentially perfect, and besides this the power of the telescope is superadded to give absolute direction, or pointing on the object under observation.

Excellent clocks are necessary for precise observation with such an instrument; but neither Tycho Brahe, nor Hevelius was provided with such accessories. Hevelius did not avail himself of the telescope as an aid to precision of observation, claiming that pinhole sights gave him more accurate results. It was a dispute concerning this question that Halley was sent over from London to Danzig to arbitrate.

There could be but one way to decide; the telescope with its added power magnifies any displacement of the instrument, and thereby enables the observer to point his instrument more exactly. So he can detect smaller errors and differences of direction than he can without it. And what is of great importance in more modern astronomy, the telescope makes it possible to observe accurately the position of objects so faint that they are wholly invisible to the naked eye.

CHAPTER X

KEPLER, THE GREAT CALCULATOR

Most fortunate it was for the later development of astronomical theory that Tycho Brahe not only was a practical or observational astronomer of the highest order, but that he confined himself studiously for years to observations of the places of the planets. Of Mars he accumulated an especially long and accurate series, and among those who assisted him in his work was a young and brilliant pupil named Johann Kepler.

Strongly impressed with the truth of the Copernican System, Kepler was free to reject the erroneous compromise system devised by Tycho Brahe, and soon after Tycho's death Kepler addressed himself seriously to the great problem that no one had ever attempted to solve, viz: to find out what the laws of motion of the planets round the sun really are. Of course he took the fullest advantage of all that Ptolemy and Copernicus had done before him, and he had in addition the splendid observations of Tycho Brahe as a basis to work upon.

Copernicus, while he had effected the tremendous advance of substituting the sun for the earth as the center of motion, nevertheless clung to the erroneous notion of Ptolemy that all the bodies of the sky must perforce move at uniform speeds, and in circular curves, the circle being the only "perfect curve." Kepler was not long in finding out that this could not be so, and he found it out because Tycho Brahe's observations were much more accurate than any that Copernicus had employed.

Naturally he attempted the nearest planet first, and that was Mars—the planet that Tycho had assigned to him for research. How fortunate that the orbit of Mars was the one, of all the planets, to show practically the greatest divergence from the ancient conditions of uniform motion in a perfectly circular orbit! Had the orbit of Mars chanced to be as nearly circular as is that of Venus, Kepler might well have been driven to abandon his search for the true curve of planetary motion.

However, the facts of the cosmos were on his side, but the calculations essential in testing his various hypotheses were of the most tedious nature, because logarithms were not yet known in his day. His first discovery was that the orbit of Mars is certainly not a circle, but oval or elliptic in figure. And the sun, he soon found, could not be in the center of the ellipse, so he made a series of trial calculations with the sun located in one of the foci of the ellipse instead.

Then he found he could make his calculated places of Mars agree quite perfectly with Tycho Brahe's observed positions, if only he gave up the other ancient requisite of perfectly uniform motion. On doing this, it soon appeared that Mars, when in perihelion, or nearest the sun, always moved swiftest, while at its greatest distance from the sun, or aphelion, its orbital velocity was slowest.

Kepler did not busy himself to inquire why these revolutionary discoveries of his were as they were; he simply went on making enough trials on Mars, and then on the other planets in turn, to satisfy himself that all the planetary orbits are elliptical, not circular in form, and are so located in space that the center of the sun is at one of the two foci of each orbit. This is known as Kepler's first law of planetary motion.

The second one did not come quite so easy; it concerned the variable speed with which the planet moves at every point of the orbit. We must remember how handicapped he was in solving this problem: only the geometry of Euclid to work with, and none of the refinements of the higher mathematics of a later day. But he finally found a very simple relation which represented the velocity of the planet everywhere in its orbit. It was this: if we calculate the area swept, or passed over, by the planet's radius vector (that is, the line joining its center to the sun's center) during a week's time near perihelion, and then calculate the similar area for a week near aphelion, or indeed for a week when Mars is in any intermediate part of its orbit, we shall find that these areas are all equal to each other. So Kepler formulated his second great law of planetary motion very simply: the radius vector of any planet describes, or sweeps over, equal areas in equal times. And he found this was true for all the planets.

But the real genius of the great mathematician was shown in the discovery of his third law, which is more complex and even more significant than the other two—a law connecting the distances of the planets from the sun with their periods of revolution about the sun. This cost Kepler many additional years of close calculation, and the resulting law, his third law of planetary motion is this: The cubes of the mean or average distances of the planets from the sun are proportional to the squares of their times of revolution around him.

So Kepler had not only disposed of the sacred theories of motion of the planets held by the ancients as inviolable, but he had demonstrated the truth of a great law which bound all the bodies of the solar system together. So accurately and completely did these three laws account for all the motions, that the science of astronomy seemed as if finished; and no matter how far in the future a time might be assigned, Kepler's laws provided the means of calculating the planet's position for that epoch as accurately as it would be possible to observe it. Kepler paused here, and he died in 1630.

CHAPTER XI

GALILEO, THE GREAT EXPERIMENTER

The fifteenth and sixteenth centuries, containing the lives and work of Copernicus, Tycho, Galileo, Kepler, Huygens, Halley, and Newton, were a veritable Golden Age of astronomy. All these men were truly great and original investigators.

None had a career more picturesque and popular than did Galileo. Born a few years earlier and dying a few years later than Kepler, the work of each of these two great astronomers was wholly independent of the other and in entirely different fields. Kepler was discovering the laws of planetary motion, while Galileo was laying the secure foundations of the new science of dynamics, in particular the laws of falling bodies, that was necessary before Kepler's laws could be fully understood. When only eighteen Galileo's keen power of observation led to his discovery of the laws of pendulum motion, suggested by the oscillation to and fro of a lamp in the cathedral of Pisa.

The world-famous leaning tower of this place, where he was born, served as a physical laboratory from the top of which he dropped various objects, and thus was led to formulate the laws of falling bodies. He proved that Aristotle was all wrong in saying that a heavy body must fall swifter in proportion to its weight than a lighter one. These and other discoveries rendered him unpopular with his associates, who christened him the "Wrangler."

The new system of Copernicus appealed to him; and when he, first of all men, turned a telescope on the heavenly bodies, there was Venus with phases like those of the moon, and Jupiter with satellites traveling about it—a Copernican system in miniature. Nothing could have happened that would have provided a better demonstration of the truth of the new system and the falsity of the old. His marvelous discoveries caused the greatest excitement—consternation even, among the anti-Copernicans. Galileo published the "Sidereus Nuncius," with many observations and drawings of the moon, which he showed to be a body not wholly dissimilar to the earth: this, too, was obviously of great moment in corroboration of the Copernican order and in contradiction to the Ptolemaic, which maintained sharp lines of demarcation between things terrestrial and things celestial.

His telescopes, small as they were, revealed to him anomalous appearances on both sides of the planet Saturn which he called *ansæ*, or handles. But their subsequent disappearance was unaccountable to him, and later observers, who kept on guessing ineffectively till Huygens, nearly a half century after, showed that the true nature of the appendage was a ring. Spots on the sun were frequently observed by Galileo and led to bitter controversies. He proved, however, that they were objects on the sun itself, not outside it, and by noticing their repeated transits across the sun's disk, he showed that the sun turned round on his axis in a little less than a month—another analogy to the like motion of the earth on the Copernican plan.

Galileo's appointment in 1610 as "First Philosopher and Mathematician" to the Grand Duke of Tuscany gave him abundant time for the pursuit of original investigations and the preparation of books and pamphlets. His first visit to Rome the year following was the occasion of a reception with great honor by many cardinals and others of high rank. His lack of sympathy with others whose views differed from his, and his naturally controversial spirit, had begun to lead him headlong into controversies with the Jesuits and the church, which culminated in his censure by the authorities of the church and persecution by the Inquisition.

In 1618 three comets appeared, and Galileo was again in controversial hot water with the Jesuits. But it led to the publication five years later of "Il Saggiatore" (The Assayer), of no great scientific value, but only a brilliant bit of controversial literature dedicated to the newly elevated Pope, Urban VIII. Later he wrote through several years a great treatise, more or less controversial in character, entitled a "Dialogue on the Two Chief Systems of the World" between three speakers, and extending through four successive days. Simplicio argues for the Aristotelians, Salviati for the Copernicans, while Sagredo does his best to be neutral. It will always be a very readable book, and we are fortunate to have a recent translation by Professor Crew of Evanston.

Here we find the first suggestion of the modern method of getting stellar parallaxes, the relative parallax, that is, of two stars in the same field—a method not put into service till Bessel's time, two centuries later. But the most important chapters of the "Dialogue" deal with Galileo's investigations of the laws of motion of bodies in general, which he applied to the problem of the earth's motion. In this he really anticipated Newton in the first of his three laws of motion, and in a subsequent work, dealing with the theory of projectiles, he reaches substantially the results of Newton's second law of motion, although he gave no general statement of the principle. Nevertheless, in the epoch where his life was lived and his work done, his telescopic discoveries, combined with his dynamic researches in untrodden fields, resulted in the complete and final overthrow of the ancient system of error, and the secure establishment of the Copernican system beyond further question and discussion. Only then could the science of astronomy proceed unhampered to the fullest development by the master minds of succeeding centuries.

CHAPTER XII

AFTER THE GREAT MASTERS

Following Kepler and Galileo was a half century of great astronomical progress along many lines laid out by the work of the great masters. The telescope seemed only a toy, but its improvement in size and quality showed almost inconceivable possibilities of celestial discoveries.

Hevelius of Danzig took up the study of the moon, and his "Selenographia" was finely illustrated by plates which he not only drew but engraved himself. Lunar names of mountains, plains, and craters we owe very largely to him. Also he published among other works two on comets, the second of which was published in 1668 and called the "Cometographia," the first detailed account of all the comets observed and recorded to date.

Many were the telescopes turned on the planet Saturn, and every variety of guess was made as to the actual shape and physical nature of the weird appendages discovered by Galileo. The true solution was finally reached by Huygens, whose mechanical genius had enabled him to grind and polish larger and better lenses than his contemporaries; in 1659 he published the "Systema Saturnium" interpreting the ring and the cause of its various configurations, and the first discovery of a Saturnian satellite is due to him.

Gascoigne in England about 1640 was the first to make the important application of the micrometer to enhance the accuracy of measurement of small angles in the telescopic field; an invention made and applied independently many years later by Huygens in Holland and Auzout and Picard in France, where the instrument was first regularly employed as an accessory in the work of an observatory.

Another Englishman, Jeremiah Horrocks, was the first observer of a transit of Venus over the disk of the sun, in 1639. Horrocks was possessed of great ability in calculational astronomy also. This was about the time of the invention of the pendulum clock by Huygens, which in conjunction with the later invention of the transit instrument by Roemer wrought a revolution in the exacting art of practical astronomy. This was because it enabled the time to be carried along continuously, and the revolution of the earth could be utilized in making precise measures of the position of sun, moon, and stars. Louis XIV had just founded the new Observatory at Paris in 1668, and Picard was the first to establish regular time-observations there.

Huygens followed up the motion of the pendulum in theory as well as practice in his "Horologium Oscillatorium" (1673), showing the way to measure the force of gravity, and his study of circular motion showed the fundamental necessity of some force directed toward the center in planetary motions.

The doctrine of the sphericity of the earth being no longer in doubt, the great advance in accuracy of astronomical observation indicated to Willebrord Snell in Holland the best way to measure an arc of meridian by triangulation. Picard repeated the measurements near Paris with even greater accuracy, and his results were of the utmost significance to Newton in establishing his law of gravitation.

Domenico Cassini, an industrious observer, voluminous writer, and a strong personality, devised telescopes of great size, discovered four Saturnian satellites and the main division in the ring of Saturn, determined the rotation periods of Mars and Jupiter, and prepared tables of the eclipses of Jupiter's satellites. At his suggestion Richer undertook an expedition to Cayenne in latitude 5 degrees north, where it was found that the intensity of gravity was less than at Paris, and his clock therefore lost time, thus indicating that the earth was not a perfect sphere as had been thought, but a spheroid instead.

The planet Mars passed a near opposition, and Richer's observations of it from Cayenne, when combined with those of Cassini and others in France, gave a new value of the sun's parallax and distance, really the first actual measurement worth the name in the history of astronomy.

To close this era of signal advance in astronomy we may cite a discovery by Roemer of the first order: no less than that of the velocity of transmission of light through space. At the instigation of Picard, Roemer in studying the motions of Jupiter's satellites found that the intervals between eclipses grew less and less as Jupiter and the earth approached each other, and greater and greater than the average as the two planets separated farther and farther. Roemer correctly attributed this difference to the progressive motion of light and a rough value of its velocity was calculated, though not accepted by astronomers generally for more than a century.

Why the laws of Kepler should be true, Kepler himself was unable to say. Nor could anyone else in that day answer these questions: (1) The planets move in orbits that are elliptical not circular—why should they move in an imperfect curve, rather than the perfect one in which it had always been taught that they moved? (2) Why should our planet vary its velocity at all, and travel now fast, now slow; especially why should the speed so vary that the line of varying length, joining the planet to the sun, always passes over areas proportional to the time of describing them? And (3) Why should there be any definite relation of the distances of planets from the sun to their times of revolution about him? Why should it be exactly as the cube of one to the square of the other?

We must remember that the Copernican system itself was not yet, in the beginning of the seventeenth century, accepted universally; and the great minds of that period were most concerned in overturning the erroneous theory of Ptolemy.

The next step in logical order was to find a basic explanation of the planetary motions, and Descartes and his theory of vortices are worthy of mention, among many unsuccessful attempts in this direction. Descartes was a brilliant French philosopher and mathematician, but his hypothesis of a multitude of whirlpools in the ether, while ingenious in theory, was too vague and indefinite to account for the planetary motions with any approach to the precision with which the laws of Kepler represented them.

Another great astronomer whose labors helped immensely in preparing the way for the signal discoveries that were soon to come was Huygens, a man of versatility as natural philosopher, mechanician, and astronomical observer. Huygens was born thirteen years before the death of Galileo, and to the discovery of the laws of motion by the latter Huygens added researches on the laws of action of centrifugal forces. Neither of them, however, appeared to see the immediate bearing on the great general problem of celestial motions in its true light, and it was reserved for another generation, and an astronomer of another country, to make the one fundamental discovery that should explain the whole by a single simple law.

CHAPTER XIII

NEWTON AND MOTION

"How is it that you are able to make these great discoveries?" was once asked of Sir Isaac Newton, *facile princeps* of all philosophers, and the discoverer of the great law of universal gravitation.

"By perpetually thinking about them," was Newton's terse and illuminating reply. He had set for himself the definite problem of Kepler's laws: why is it that they are true, and is there not some single, general law that will embody all the circumstances of the planetary motions?

Newton was born in 1643, the year after the death of Galileo. He had a thorough training in the mathematics of his day, and addressed himself first to an investigation and definite formulation of the general laws of motion, which he found to be three in number, and which he was able to put in very simple terms. The first one is: Any body, once it is set in motion, will continue to move forward in a straight line with a uniform velocity forever, provided it is acted upon by no force whatever. In other words, a state of motion is as natural as a state of rest (rest in relation to things everywhere adjacent) in which we find all things in general.

Here on earth where gravity itself pulls all objects downward toward the earth, and where resistance of the air tends to hold a moving body back and bring it to rest, and where friction from contact with whatever material substance may be in its path is perpetually tending to neutralize all motion—with all three of these forces always at work to stop a moving body, the truth of this first and fundamental law of motion was not apparent on the surface.

Till Galileo's time everyone had made the mistake of supposing that some force or other must be acting continually on every moving body to keep it in motion. Ptolemy, Copernicus, Kepler, Leonardo da Vinci—all failed to see the truth of this law which Newton developed in the immortal *Principia*. And at the present day it is not always easy to accept at first, although the progress of mechanical science, by reducing friction and resistance, has produced machines in which motion of large masses may be kept up indefinitely with the application of only the merest minimum of force.

Once a planet is set in motion round the sun, it would go on forever through frictionless, non-resistant space; but there must be a central force, as Huygens saw clearly, to hold it in its orbit. Otherwise it would at any moment take the direction of a tangent to the orbit. Here is where Newton's second law of motion comes in, and he formulated it with great definiteness. When any force acts on a moving body, its deviation from a straight line will be in the direction of the force applied and proportional to that force.

In accord with this law, Newton first began to inquire whether the force of attraction here on earth, which everyone commonly recognizes as gravity, drawing all things down toward the center of the earth, might not extend upward indefinitely. It is found in operation on the summits of mountain peaks, and the clouds above them and the rain falling from them are obviously drawn downward by the same force. May it not extend outward into space, even as far as the moon?

This was an audacious question, but Newton not only asked, but tried to answer it in the year 1665, when he was only twenty-three. On the surface of the earth this attraction is strong enough to draw a falling body downward through a vertical space of sixteen feet in a second of time. What ought it to be at the distance of the moon. The distance of the moon in Newton's time was better known in terms of the earth's size than was the size of the earth itself: the earth's radius was known to be one-sixtieth of the moon's distance, but the earth's diameter was thought to be something under 7,000 miles, so that Newton's first calculations were most disappointing, and he laid them aside for nearly twenty years.

Meanwhile the French astronomers led by Picard had measured the earth anew, and showed it to be nearly 8,000 miles in diameter. As soon as Newton learned of this, he revised his calculations, and found that by the law of the inverse square the moon, in one second, should fall away from a tangent to its orbit one thirty-six hundredth of sixteen feet.

This accorded exactly with his original supposition that the earth's attraction extended to the moon. So he concluded that the force which makes a stone fall, or an apple, as the story goes, is the same force that holds the moon in its orbit, and that this force diminishes in the exact proportion that the square of the distance from the earth's center increases. The moon, indeed, becomes a falling body; only, as Kingdon Clifford puts it: "She is going so fast and is so far off that she falls quite around to the other side of the earth, instead of hitting it; and so goes on forever."

Newton goes on in the *Principia* to explain the extension of gravitation to the other bodies of the solar system beyond the earth and moon. Clearly the same gravitation that holds the moon in its orbit round the earth, must extend outward from the sun also, and hold all the planets in their orbits centered about him. Newton demonstrates by calculation based on Kepler's third law that (1) the forces drawing the planets toward the sun are inversely as the squares of their mean distances from him; and (2) if the force be constantly directed toward the sun, the radius vector in an elliptic orbit must pass over equal areas in equal times.

Nicholas Copernicus

Galileo Galilei

Johann Kepler

Sir Isaac Newton

CHAPTER XIV
NEWTON AND GRAVITATION

So all of Kepler's laws could be embodied in a single law of gravitation toward a central body, whose force of attraction decreases outward in exact proportion as the square of the distance increases.

Only one farther step had to be taken, and this the most complicated of all: he must make all the bodies of the sky conform to his third law of motion. This is: Action and reaction are equal, or the mutual actions of any two bodies are always equal and oppositely directed. There must be mutual attractions everywhere: earth for sun as well as sun for earth, moon for sun and sun for moon, earth for Venus and Venus for earth, Jupiter for Saturn and Saturn for Jupiter, and so on.

The motions of the planets in the undisturbed ellipses of Kepler must be impossible. As observations of the planets became more accurate, it was found that they really did fail to move in exact accord with Kepler's laws unmodified. Newton was unable, with the imperfect processes of the mathematics of his day to ascertain whether the deviations then known could be accounted for by his law of gravitation; but he nevertheless formulated the law with entire precision, as follows:

Every particle of matter in the universe attracts every other particle with a force exactly proportioned to the product of their masses, and inversely as the square of the distance between their centers.

The centuries of astronomical research since Newton's day, however, have verified the great law with the utmost exactness. Practically every irregularity of lunar and planetary motion is accounted for; indeed, the intricacies of the problems involved, and the nicety of their solution, have led to the invention of new mathematical processes adequate to the difficulties encountered.

And about the middle of the last century, when Uranus departed from the path laid out for it by the mathematical astronomers, its orbital deviations were made the basis of an investigation which soon led to the assignment of the position where a great planet could be found that would account for the unexplained irregularities of the motion of Uranus. And the immediate discovery of this planet, Neptune, became the most striking verification of the Newtonian law that the solar system could possibly afford.

The astronomers of still later days investigating the statelier motions of stellar systems find the Newtonian law regnant everywhere among the stars where our most powerful telescopes have as yet reached. So that Newton's law is known as the law of Universal Gravitation, and its author is everywhere held as the greatest scientist of the ages.

Newton's *Principia* may be regarded as the culminating research of the inductive method, and further outline of its contents is desirable. It is divided into three books following certain introductory sections. The first book treats of the problems of moving bodies, the solutions being worked out generally and not with special reference to astronomy. The second book deals with the motion of bodies through resistant media, as fluids, and has very little significance in astronomy. The third book is the all important one, and applies his general principles to the case of the actual solar system, providing a full explanation of the motions of all the bodies of the system known in his day. Anyone who critically reads the *Principia* of Newton will be forced to conclude that its author was a genius in the highest sense of the word. The elegance and thoroughness of the demonstrations, and the completeness of application of the law of gravitation are especially impressive.

The universality of his new law was the feature to which he gave particular attention. It was clear to him that the gravitation of a planet, although it acted as if wholly concentrated at the center, was nevertheless resident in every one of the particles of which the planet is composed. Indeed, his universal law was so formulated as to make every particle attract every other particle; and an investigation known as the Cavendish experiment—a research of great delicacy of manipulation—not only proves this, but leads also to a measurement of the earth's mean density, from which we can calculate approximately how much the earth actually weighs.

Another way to attack the same problem is by measuring the attraction of mountains, as Maskelyne, Astronomer Royal of Scotland did on Mount Schehallien in Scotland, which was selected because of its sheer isolation. The attraction of the mountain deflected the plumb-lines by measurable amounts, the volume of the mountain was carefully ascertained by surveys, and geologists found out what rocks composed it. So the weight of the entire mountain became pretty well known, and combining this with the observed deflection, an independent value of the earth's weight was found.

Still other methods have been applied to this question, and as an average it is found that the materials composing the earth are about five and a half times as heavy as water, and the total weight of the earth is something like six sextillions of tons.

What is the true shape of the earth? And does the earth's turning round on its axis affect this shape? Newton saw the answer to these questions in his law of gravitation. A spherical figure followed as a matter of course from the mutual attraction of all materials composing the earth, providing it was at rest, or did not turn round on its axis. But rotation bulges it at the equator and draws it in at the poles, by an amount which calculation shows to be in exact agreement with the amount ascertained by actual measurement of the earth itself.

Another curious effect, not at first apparent, was that all bodies carried from high latitudes toward the equator would get lighter and lighter, in consequence of the centrifugal force of rotation. This was unexpectedly demonstrated by Richer when the French Academy sent him south to observe Mars in 1672. His clock had been regulated exactly in Paris, and he soon found that it lost time when set up at Cayenne. The amount of loss was found by observation, and it was exactly equal to the calculated effect that the reduction of gravity by centrifugal action should produce.

Also Newton saw that his law of gravitation would afford an explanation of the rise and fall of the tides. The water on the side of the earth toward the moon, being nearer to the moon, would be more strongly attracted toward it, and therefore raised in a tide. And the water on the farther side of the earth away from the moon, being at a greater distance than the earth itself, the moon would attract the earth more strongly than this mass of water, tending therefore to draw the earth away from the water, and so raising at the same time a high tide on the side of the earth away from the moon. As the earth turns round on its axis, therefore, two tidal waves continually follow each other at intervals of about twelve hours.

The sun, too, joins its gravitating force with that of the moon, raising tides nearly half as high as those which the moon produces, because the sun's vaster mass makes up in large part for its much greater distance. At first and third quarters of the moon, the sun acts against the moon, and the difference of their tide-producing forces gives us "neap tides"; while at new moon and full, sun and moon act together, and produce the maximum effect known as "spring tides."

Newton passed on to explain, by the action of gravitation also, the precession of the equinoxes, a phenomenon of the sky discovered by Hipparchus, who pretty well ascertained its amount, although no reason for it had ever been assigned. The plane of the earth's equator extended to the celestial sphere marks out the celestial equator, and the two opposite points where it intersects the plane of the ecliptic, or the earth's path round the sun, are called the equinoctial points, or simply the equinoxes. And precession of the equinoxes is the motion of these points westward or backward, about 50 seconds each year, so that a complete revolution round the ecliptic would take place in about 26,000 years.

Newton saw clearly how to explain this: it is simply due to the attraction of the sun's gravitation upon the protuberant bulge around the earth's equator, acting in conjunction with the earth's rotation on its axis, the effect being very similar to that often seen in a spinning top, or in a gyroscope. The moon moving near the ecliptic produces a precessional effect, as also do the planets to a very slight degree; and the observed value of precession is the same as that calculated from gravitation, to a high degree of precision.

Newton died in 1727, too early to have witnessed that complete and triumphant verification of his law which ultimately has accounted for practically every inequality in the planetary motions caused by their mutual attractions. The problems involved are far beyond the complexity of those which the mathematical astronomer has to deal with, and the mathematicians of France deserve the highest credit for improving the processes of their science so that obstacles which appeared insuperable were one after another overcome.

Newton's method of dealing with these problems was mainly geometric, and the insufficiency of this method was apparent. Only when the French mathematicians began to apply the higher methods of algebra was progress toward the ultimate goal assured. D'Alembert and Clairaut for a time were foremost in these researches, but their places were soon taken by Lagrange, who wrote the "Mécanique Analytique," and Laplace, whose "Mécanique Céleste" is the most celebrated work of all. In large part these works are the basis of the researches of subsequent mathematical astronomers who, strictly speaking, cannot as yet be said to have arrived at a complete and rigorous solution of all the problems which the mutual attractions of all the bodies of the solar system have originated.

It may well be that even the mathematics of the present day are incompetent to this purpose. When the brilliant genius of Sir William Hamilton invented quaternion analysis and showed the marvelous facility with which it solved the intricate problems of physics, there was the expectation that its application to the higher problems of mathematical astronomy might effect still greater advances; but nothing in that direction has so far eventuated. Some astronomers look for the invention of new functions with numerical tables bearing perhaps somewhat the relation to present tables of logarithms, sines, tangents, and so on, that these tables do to the simple multiplication table of Pythagoras.

CHAPTER XV

AFTER NEWTON

We have said that practically all the motions in the solar system have been accounted for by the Newtonian law of gravitation. It will be of interest to inquire into the instances that lead to qualification of this absolute statement.

One relates to the planet Mercury, whose orbit or path round the sun is the most elliptical of all the planetary orbits. This will be explained a little later.

The moon has given the mathematical astronomers more trouble than any other of the celestial bodies, for one reason because it is nearest to us and very minute deviations in its motion are therefore detectible. Halley it was who ascertained two centuries ago that the moon's motion round the earth was not uniform, but subject to a slight acceleration which greatly puzzled Lagrange and Laplace, because they had proved exactly this sort of thing to be impossible, unless indeed the body in question should be acted on by some other force than gravitation. But Laplace finally traced the cause to the secular or very slow reduction in the eccentricity of the earth's own orbit. The sun's action on the moon was indeed progressively changing from century to century in such manner as to accelerate the moon's own motion in its orbit round the earth.

Adams, the eminent English astronomer, revised the calculations of Laplace, and found the effect in question only half as great as Laplace had done; and for years a great mathematical battle was on between the greatest of astronomical experts in this field of research. Adams, in conjunction with Delaunay, the greatest of the French mathematicians a half century ago, won the battle in so far as the mathematical calculations were concerned; but the moon continues to the present day her slight and perplexing deviation, as if perhaps our standard time-keeper, the earth, by its rotation round its axis, were itself subject to variation. Although many investigations have been made of the uniformity of the earth's rotation, no such irregularity has been detected, and this unexplained variation of the moon's motion is one of the unsolved problems of the gravitational astronomer of to-day.

But we are passing over the most impressive of all the earlier researches of Lagrange and Laplace, which concerned the exceedingly slow changes, technically called the secular variations of the elements of the planetary orbits. These elements are geometrical relations which indicate the form of the orbit, the size of the orbit, and its position in space; and it was found that none of these relations or quantities are constant in amount or direction, but that all, with but one exception, are subject to very slow, or secular, change, or oscillation.

This question assumed an alarming significance at an early day, particularly as it affected the eccentricity of the earth's orbit round the sun. Should it be possible for this element to go on increasing for indefinite ages, clearly the earth's orbit would become more and more elliptical, and the sun would come nearer and nearer at perihelion, and the earth would drift farther and farther from the sun at aphelion, until the extremes of temperature would bring all forms of life on the earth to an end. The refined and powerful analysis of Lagrange, however, soon allayed the fears of humanity by accounting for these slow progressive changes as merely part of the regular system of mere oscillations, in entire accord with the operation of the law of gravitation; and extending throughout the entire planetary system. Indeed, the periods of these oscillations were so vast that none of them were shorter than 50,000 years, while they ranged up to two million years in length—"great clocks of eternity which beat ages as ours beat seconds."

About a century ago, an eminent lecturer on astronomy told his audience that the problem of weighing the planets might readily be one that would seem wholly impossible to solve. To measure their sizes and distances might well be done, but actually to ascertain how many tons they weigh—never!

Yet if a planet is fortunate enough to have one satellite or more, the astronomer's method of weighing the planet is exceedingly simple; and all the major planets have satellites except the two interior ones, Mercury and Venus. As the satellite travels round its primary, just as the moon does round the earth, two elements of its orbit need to be ascertained, and only two. First, the mean distance of the satellite from its primary, and second the time of revolution round it.

Now it is simply a case of applying Kepler's third law. First take the cube of the satellite's distance and divide it by the square of the time of revolution. Similarly take the cube of the planet's distance from the sun and divide by the square of the planet's time of revolution round him. The proportion, then, of the first quotient to the second shows the relation of the mass (that is the weight) of the planet to that of the sun. In the case of Jupiter, we should find it to be 1,050, in that of Saturn 3,500, and so on.

The range of planetary masses, in fact, is very curious, and is doubtless of much significance in the cosmogony, with which we deal later. If we consider the sun and his eight planets, the mass or weight of each of the nine bodies far exceeds the combined mass of all the others which are lighter than itself.

To illustrate: suppose we take as our unit of weight the one-billionth part of the sun's weight; then the planets in the order of their masses will be Mercury, Mars, Venus, Earth, Uranus, Neptune, Saturn, and Jupiter. According to their relative masses, then, Mercury being a five-millionth part the weight of the sun will be represented by 200; similarly Venus, a four hundred and twenty-five thousandth part by 2,350, and so on. Then we have

Mercury	200
Mars	340
Sum of weights of Mercury and Mars	540

Venus	2,350
Sum of weights of Mercury, Mars, and Venus	2,890
The Earth	3,060
Sum of weights of four inner planets	5,950
Uranus	44,250
Sum of weights of five planets	50,200
Neptune	51,600
Sum of weights of six planets	101,800
Saturn	285,580
Sum of weights of seven planets	387,380
Jupiter	954,300
Sum of weights of all the planets	1,341,680
Mass or weight of the sun	1,000,000,000

Curious and interesting it is that Saturn is nearly three times as heavy as the six lighter planets taken together, Jupiter between two and three times heavier than all the other planets combined, while the sun's mass is 750 times that of all the great planets of his system rolled into one.

All the foregoing masses, except those of Mercury and Venus, are pretty accurately known because they were found by the satellite method just indicated. Mercury's mass is found by its disturbing effects on Encke's comet whenever it approaches very near. The mass of Venus is ascertained by the perturbations in the orbital motion of the earth. In such cases the Newtonian law of gravitation forms the basis of the intricate and tedious calculations necessary to find out the mass by this indirect method.

Its inferiority to the satellite method was strikingly shown at the Observatory in Washington soon after the satellites of Mars were discovered in 1877. The inaccurate mass of that planet, as previously known by months of computation based upon years and years of observation, was immediately discarded in favor of the new mass derived from the distance and period of the outer satellite by only a few minutes' calculation.

In weighing the planets, astronomers always use the sun as the unit. What then is the sun's own weight? Obviously the law of gravitation answers this question, if we compare the sun's attraction with the earth's at equal distances. First we conceive of the sun's mass as if all compressed into a globe the size of the earth, and calculate how far a body at the surface of this globe would fall in one second. The relation of this number to 16.1 feet, the distance a body falls in one second on the actual earth, is about 330,000, which is therefore the number of times the sun's weight exceeds that of the earth.

A word may be added regarding the force of gravitation and what it really is. As a matter of fact Newton did not concern himself in the least with this inquiry, and says so very definitely. What he did was to discover the law according to which gravitation acts everywhere throughout the solar system. And although many physicists have endeavored to find out what gravitation really is, its cause is not yet known. In some manner as yet mysterious it acts instantaneously over distances great and small alike, and no substance has been found which, if we interpose it between two bodies, has in any degree the effect of interrupting their gravitational tendency toward each other.

While the Newtonian law of gravitation has been accepted as true because it explained and accounted for all the motions of the heavenly bodies, even including such motions of the stars as have been subjected to observation, astronomers have for a long time recognized that quite possibly the law might not be absolutely exact in a mathematical sense, and that deviations from it would surely make their appearance in time.

A crude instance of this was suggested about a century ago, when the planet Uranus was found to be deviating from the path marked out for it by Bouvard's tables based on the Newtonian law; and the theory was advocated by many astronomers that this law, while operant at the medium distances from the sun where the planets within Jupiter and Saturn travel, could not be expected to hold absolutely true at the vast distance of Uranus and beyond. The discovery of Neptune in 1846, however, put an end to all such speculation, and has universally been regarded as an extraordinary verification of the law, as indeed it is.

When, however, Le Verrier investigated the orbit of Mercury he found an excess of motion in the perihelion point of the planet's orbit which neither he nor subsequent investigators have been able to account for by Newtonian gravitation, pure and simple. If Newton's theory is absolutely true, the excess motion of Mercury's perihelion remains a mystery.

Only one theory has been advanced to account for this discrepancy, and that is the Einstein theory of gravitation. This ingenious speculation was first propounded in comprehensive form nearly fifteen years ago, and its author has developed from it mathematical formulæ which appear to yield results even more precise than those based on the Newtonian theory.

In expressing the difference between the law of gravitation and his own conception, Einstein says: "Imagine the earth removed, and in its place suspended a box as big as a moon or a whole house and inside a man naturally floating in the center, there being no force whatever pulling him. Imagine, further, this box being, by a rope or other contrivance, suddenly jerked to one side, which is scientifically termed 'difform motion,' as opposed to 'uniform motion.' The person would then naturally reach bottom on the opposite side. The result would consequently be the same as if he obeyed Newton's law of gravitation, while, in fact, there is no gravitation exerted whatever, which proves that difform motion will in every case produce the same effects as gravitation…. The term relativity refers to time and space. According to Galileo and Newton, time and space were absolute entities, and the moving systems of the universe were dependent on this absolute time and space. On this conception was built the science of mechanics. The resulting formulas sufficed for all motions of a slow nature; it was found, however, that they would not conform to the rapid motions apparent in electrodynamics.... Briefly the theory of special relativity discards absolute time and space, and makes them in every instance relative to moving systems. By this theory all phenomena in electrodynamics, as well as mechanics, hitherto irreducible by the old formulæ, were satisfactorily explained."

Natural phenomena, then, involving gravitation and inertia, as in the planetary motions, and electro-magnetic phenomena, including the motion of light, are to be regarded as interrelated, and not independent of one another. And the Einstein theory would appear to have received a striking verification in both these fields. On this theory the Newtonian dynamics fails when the velocities concerned are a near approach to that of light. The Newtonian theory, then, is not to be considered as wrong, but in the light of a first approximation. Applying the new theory to the case of the motion of Mercury's perihelion, it is found to account for the excess quite exactly.

On the electro-magnetic side, including also the motion of light, a total eclipse of the sun affords an especially favorable occasion for applying the critical test, whether a huge mass like the sun would or would not deflect toward itself the rays of light from stars passing close to the edge of its disk, or limb. A total eclipse of exceptional duration occurred on May 29, 1919, and the two eclipse parties sent out by the Royal Society of London and the Royal Astronomical Society were equipped especially with apparatus for making this test. Their stations were one on the east coast of Brazil and the other on the west coast of Africa.

Accurate calculation beforehand showed just where the sun would be among the stars at the time of the eclipse; so that star plates of this region were taken in England before the expeditions went out. Then, during the total eclipse, the same regions were photographed with the eclipsed sun and the corona projected against them. To make doubly sure, the stars were a third time photographed some weeks after the eclipse, when the sun had moved away from that particular region.

Measuring up the three sets of plates, it was found that an appreciable deflection of the light of the stars nearest alongside the sun actually exists; and the amount of it is such as to afford a fair though not absolutely exact verification of the theory. The observed deflection may of course be due to other causes, but the English astronomers generally regard the near verification as a triumph for the Einstein theory. Astronomers are already beginning preparations for a repetition of the eclipse programme with all possible refinement of observation, when the next total eclipse of the sun occurs, September 20, 1922, visible in Australia and the islands of the Indian Ocean.

A third test of the theory is perhaps more critical than either of the others, and this necessitates a displacement of spectral lines in a gravitational field toward the red end of the spectrum; but the experts who have so far made measures for detecting such displacement disagree as to its actual existence. The work of St. John at Mt. Wilson is unfavorable to the theory, as is that of Evershed of Kodiakanal, who has made repeated tests on the spectrum of Venus, as well as in the cyanogen bands of the sun.

The enthusiastic advocates of the Einstein theory hold that, as Newton proved the three laws of Kepler to be special cases of his general law, so the "universal relativity theory" will enable eventually the Newtonian law to be deduced from the Einstein theory. "This is the way we go on in science, as in everything else," wrote Sir George Airy, Astronomer Royal; "we have to make out that something is true; then we find out under certain circumstances that it is not quite true; and then we have to consider and find out how the departure can be explained." Meanwhile, the prudent person keeps the open mind.

CHAPTER XVI
HALLEY AND HIS COMET

Halley is one of the most picturesque characters in all astronomical history. Next to Newton himself he was most intimately concerned in giving the Newtonian law to the world.

Edmund Halley was born (1656) in stirring times. Charles I. had just been executed, and it was the era of Cromwell's Lord Protectorate and the wars with Spain and Holland. Then followed (1660) the promising but profligate Charles II. (who nevertheless founded at Greenwich the greatest of all observatories when Halley was nineteen), the frightful ravages of the Black Plague, the tyrannies of James II., and the Revolution of 1688—all in the early manhood of Halley, whose scientific life and works marched with much of the vigor of the contending personalities of state.

The telescope had been invented a half century earlier, and Galileo's discoveries of Jupiter's moons and the phases of Venus had firmly established the sun-centered theory of Copernicus.

The sun's distance, though, was known but crudely; and why the stars seemed to have no yearly orbits of their own corresponding to that of the earth was a puzzle. Newton was well advanced toward his supreme discovery of the law of universal gravitation; and the authority of Kepler taught that comets travel helter-skelter through space in straight lines past the earth, a perpetual menace to humanity.

"Ugly monsters," that comets always were to the ancient world, the medieval church perpetuated this misconception so vigorously that even now these harmless, gauzy visitors from interstellar space possess a certain "wizard hold upon our imagination." This entertaining phase of the subject is excellently treated in President Andrew D. White's "History of the Doctrine of Comets," in the Papers of the American Historical Association. Halley's brilliant comet at its earlier apparitions had been no exception.

Halley's father was a wealthy London soap maker, who took great pride in the growing intellectuality of his son. Graduating at Queen's College, Oxford, the latter began his astronomical labors at twenty by publishing a work on planetary orbits; and the next year he voyaged to St. Helena to catalogue the stars of the southern firmament, to measure the force of terrestrial gravity, and observe a transit of Mercury over the disk of the sun.

While clouds seriously interfered with his observations on that lonely isle, what he saw of the transit led to his invention of "Halley's method," which, as applied to the transit of Venus, though not till long after his death, helped greatly in the accurate determination of the sun's distance from the earth. Halley's researches on the proper motions of the stars of both hemispheres soon made him famous, and it was said of him, "If any star gets displaced on the globe, Halley will presently find it out."

His return to London and election to the Royal Society (of which he was many years secretary) added much to his fame, and he was commissioned by the society to visit Danzig and arbitrate an astronomical controversy between Hooke and Hevelius, both his seniors by a generation.

On the continent he associated with other great astronomers, especially Cassini, who had already found three Saturnian moons; and it was then he observed the great comet of 1680, which led up to the most famous event of Halley's life.

The seerlike Seneca may almost be said to have predicted the advent of Halley, when he wrote ("Quaestiones Naturales," vii): "Some day there will arise a man who will demonstrate in what region of the heavens comets pursue their way; why they travel apart from the planets; and what their sizes and constitution are. Then posterity will be amazed that simple things of this sort were not explained before."

To Newton it appeared probable that cometary voyagers through space might have orbits of their own; and he proved that the comet of 1680 never swerved from such a path. As it could nowhere approach within the moon's orbit, clearly threats of its wrecking the earth and punishing its inhabitants ought to frighten no more.

Halley then became intensely interested in comets, and gathered whatever data concerning the paths of all these bodies he could find. His first great discovery was that the comets seen in 1531 by Apian, and in 1607 by Kepler, traveled round the sun in identical paths with one he had himself observed in 1682. A still earlier appearance of Halley's comet (1456) seems to have given rise to a popular and long-reiterated myth of a papal bull excommunicating "the Devil, the Turk, and the Comet."

No longer room for doubt: so certain was Halley that all three were one and the same comet, completing the round of its orbit in about seventy-six years, that he fearlessly predicted that it would be seen again in 1758 or 1759. And with equal confidence he might have foretold its return in 1835 and 1910; for all three predictions have come true to the letter.

Halley's span of existence did not permit his living to see even the first of these now historic verifications. But we in our day may emphatically term the epoch of the third verified return *Annus Halleianus.*

Says Turner, Halley's successor in the Savilian chair at Oxford to-day: "There can be no more complete or more sensational proof of a scientific law, than to predict events by means of it. Halley was deservedly the first to perform this great service for Newton's Law of Gravitation, and he would have rejoiced to think how conspicuous a part England was to play in the subsequent prediction of the existence of Neptune."

Halley rose rapidly among the chief astronomical figures of his day. But he had little veneration for mere authority, and the significant veering of his religious views toward heterodoxy was for years an obstacle to his advance.

Still Halley the astronomer was great enough to question any contemporary dicta that seemed to rest on authority alone. Everyone called the stars "fixed" stars; but Halley doubting this, made the first discovery of a star's individual motion — proper motion, as astronomers say. To-day, two hundred years after, every star is considered to be in motion, and astronomers are ascertaining their real motions in the celestial spaces to a nicety undreamed of by even the exacting Halley.

The moon, of priceless service to the early navigator, was regarded by all astronomers as endowed with an average rate of motion round the earth that did not vary from age to age. But Halley questioned this too; and on comparing with the ancient value from Chaldean eclipses, he made another discovery — the secular acceleration of the moon's mean motion, as it is technically termed. This was a colossal discovery in celestial dynamics; and the reason underlying it lay hidden in Newton's law for yet another century, till the keener mathematics of Laplace detected its true origin.

With Newton, Halley laid down the firm foundations of celestial mechanics, and they pushed the science as far as the mathematics of their day would permit. Halley, however, was not content with elucidating the motion of bodies nearest the earth, and pressed to the utmost confines of the solar system known to him. Here, too, he made a signal discovery of that mutual disturbance of the planets in their motion round the sun, called the great inequality of Jupiter and Saturn.

Halley's versatile genius attacked all the great problems of the day. His observation of the sun's total eclipse in 1715 is the earliest reliable account of such a phenomenon by a trained astronomer. He described the corona minutely and was the first to see that other interesting phenomenon which only an alert observer can detect, which a great astronomer of a later day compared to the "ignition of a fine train of gunpowder," and which has ever since borne the name of "Baily's beads."

Besides being a great astronomer, Halley was a man of affairs as well, which Newton, although the greater mathematician, was not. Without Halley, Newton's superb discovery might easily have been lost to the age and nation, for the latter was bent merely on making discoveries, and on speculative contemplation of them, with never a thought of publishing to the world.

Halley, more practical and businesslike, insisted on careful writing out and publication. Newton was then only forty-two, and Halley fully fourteen years his junior. But the philosophers of that day were keenly alive to the mystery of Kepler's laws, and Halley was fully conscious of the grandeur and far-reaching significance of Newton's great generalization which embodied all three of Kepler's laws in one.

Newton at last yielded, though reluctantly, and the "Principia" was given to the world, though wholly at Halley's private charges.

But Halley was far from being completely engrossed with the absorbing problems of the sky; things terrestrial held for years his undivided attention. Imagine present-day Lords Commissioners of the Admiralty intrusting a ship of the British navy to civilian command. Yet such was their confidence in Halley that he was commissioned as captain of H. M.'s pink *Paramour* in 1698, with instructions to proceed to southern seas for geographical discoveries, and for improving knowledge of the longitude problem, and of the variations of the compass. Trade winds and monsoons, charts of magnetic variation, tides and surveys of the Channel coast, and experiments with diving bells were practical activities that occupied his attention.

Halley in 1720 became Astronomer Royal. He was the second incumbent of this great office, but the first to supply the Royal Observatory with instruments of its own, some of which adorn its walls even to-day. His long series of lunar observations and his magnetic researches were of immense practical value in navigation.

Halley lived to a ripe old age and left the world vastly better than he found it. His rise from humblest obscurity was most remarkable, and he lived to gratify all the ambitions of his early manhood. "Of attractive appearance, pleasing manners, and ready wit," says one of his biographers, "loyal, generous, and free from self-seeking, he was one of the most personally engaging men who ever held the office of Astronomer Royal."

He died in office at Greenwich in 1742.

"Halley was buried," says Chambers, "in the churchyard of St. Margaret's, Lee, not far from Greenwich, and it has lately been announced that the Admiralty have decided to repair his tomb at the public expense, no descendants of his being known." There is no suitable monument in England to the memory of one of her greatest scientific men. In any event the collection and republication of his epoch-making papers would be welcomed by astronomers of every nation.

CHAPTER XVII

BRADLEY AND ABERRATION

Living at Kew in London early in the 18th century was an enthusiastic young astronomer, James Bradley. He is famous chiefly for his accurate observations of star places which have been invaluable to astronomers of later epochs in ascertaining the proper motions of stars.

The latitude of Bradley's house in Kew was very nearly the same as the declination of the bright star Gamma Draconis, so that it passed through his zenith once every day. Bradley had a zenith sector, and with this he observed with the greatest care the zenith distance of Gamma Draconis at every possible opportunity. This he did by pointing the telescope on the star and then recording the small angle of its inclination to a fine plumb line. So accurate were his measures that he was probably certain of the star's position to the nearest second of arc.

What he hoped to find was the star's motion round a very slight orbit once each year, and due to the earth's motion in its orbit round the sun. In other words, he sought to find the star's parallax if it turned out to be a measurable quantity.

It is just as well now that his method of observation proved insufficiently delicate to reveal the parallax of Gamma Draconis; but his assiduity in observation led him to an unexpected discovery of greater moment at that time. What he really found was that the star had a regular annual orbit; but wholly different from what he expected, and very much larger in amount. This result was most puzzling to Bradley. The law of relative motion would require that the star's motion in its expected orbit should be opposite to that of the earth in its annual orbit; instead of which the star was all the time at right angles to the earth's motion.

Bradley was a frequent traveler by boat on the Thames, and the apparent change in the direction of the wind when the boat was in motion is said to have suggested to him what caused the displacement of Gamma Draconis. The progressive motion of light had been roughly ascertained by Roemer: let that be the velocity of the wind. And the earth's motion in its orbit round the sun, let that be the speed of the boat. Then as the wind (to an observer on the moving boat) always seems to come from a point in advance of the point it actually proceeds from (to an observer at rest), so the star should be constantly thrown forward by an angle given by the relation of the velocity of light to the speed of the earth in orbital revolution round the sun.

The apparent places of all stars are affected in this manner, and this displacement is called the aberration of light. Astronomers since Bradley's discovery of aberration in 1726 have devoted a great deal of attention to this astronomical constant, as it is called, and the arc value of it is very nearly 20".5. This means that light travels more than ten thousand times as fast as the earth in its orbit (186,330 miles per second as against the earth's 18.5). And we can ascertain the sun's distance by aberration also because the exact values of the velocity of light and of the constant of aberration when properly combined give the exact orbital speed of the earth; and this furnishes directly by geometry the radius of the earth's orbit, that is the distance of the sun.

In fact, this is one of the more accurate modern methods of ascertaining the distance of the sun. As early as 1880 it enabled the writer to calculate the sun's parallax equal to 8".80, a value absolutely identical with that adopted by the Paris Conference of 1896, and now universally accepted as the standard.

In whatever part of the sky we observe, every star is affected by aberration. At the poles of the ecliptic, 23½ degrees from the earth's poles, the annual aberration orbits of the stars are very small circles, 41" in diameter. Toward the ecliptic the aberration orbits become more and more oval, ellipses in fact of greater and greater eccentricity, but with their major axes all of the same length, until we reach the ecliptic itself; and then the ellipse is flattened into a straight line 41" in length, in which the star travels forth and back once a year. Exact correspondence of the aberration ellipses of the stars with the annual motion of the earth round the sun affords indisputable proof of this motion, and as every star partakes of the movement, this proof of our motion round the sun becomes many million-fold.

Indeed, if we were to push a little farther the refinement of our analysis of the effect of aberration on stellar positions, we could prove also the rotation of the earth on its axis, because that motion is swift enough to bear an appreciable ratio to the velocity of light. Diurnal aberration is the term applied to this slight effect, and as every star partakes of it, demonstration of the earth's turning round on its axis becomes many million-fold also.

CHAPTER XVIII
THE TELESCOPE

Had anyone told Ptolemy that his earth-centered system of sun, moon, and stars would ultimately be overthrown, not by philosophy but by the overwhelming evidence furnished by a little optical instrument which so aided the human eye that it could actually see systems of bodies in revolution round each other in the sky, he would no doubt have vehemently denied that any such thing was possible. To be sure, it took fourteen centuries to bring this about, and the discovery even then was without much doubt due to accident.

Through all this long period when astronomy may be said to have merely existed, practically without any forward step or development, its devotees were unequipped with the sort of instruments which were requisite to make the advance possible. There were astrolabes and armillary spheres, with crudely divided circles, and the excellent work done with them only shows the genius of many of the early astronomers who had nothing better to work with. Regarding star-places made with instruments fixed in the meridian, Bessel, often called the father of practical astronomy, used to say that, even if you provided a bad observer with the best of instruments, a genius could surpass him with a gun barrel and a cart wheel.

Before the days of telescopes, that is, prior to the seventeenth century, it was not known whether any of the planets except the earth had a moon or not; consequently the masses of these planets were but very imperfectly ascertained; the phases of Mercury and Venus were merely conjectured; what were the actual dimensions of the planets could only be guessed at; the approximate distances of sun, moon, and planets were little better than guesses; the distances of the stars were wildly inaccurate; and the positions of the stars on the celestial sphere, and of sun, moon, and planets among them were far removed from modern standards of precision—all because the telescope had not yet become available as an optical adjunct to increase the power of the human eye and enable it to see as if distances were in considerable measure annihilated.

Galileo almost universally is said to have been the inventor of the telescope, but intimate research into the question would appear to give the honor of that original invention to another, in another country. What Galileo deserves the highest praise for, however, is the reinvention independently of an "optick tube" by which he could bring distant objects apparently much nearer to him; and being an astronomer, he was by universal acknowledgment first of all men to turn a telescope on the heavenly bodies. This was in the year 1609, and his first discovery was the phase of Venus, his second the four Medicean moons or satellites of Jupiter, discoveries which at that epoch were of the highest significance in establishing the truth of the Copernican system beyond the shadow of doubt.

But the first telescopes of which we have record were made, so far as can now be ascertained, in Holland very early in the 17th century. Metius, a professor of mathematics, and Jansen and Lipperhey, who were opticians in Middelburg—all three are entitled to consideration as claimants of the original invention of the telescope. But that such an instrument was pretty well known would appear to be shown by his government's refusal of a patent to Lipperhey in 1608; while the officials recognizing the value of such an instrument for purposes of war, got him to construct several telescopes and ordered him to keep the invention a secret.

Within a year Galileo heard that an instrument was in use in Holland by which it was possible to see distant objects as if near at hand. Skilled in optics as he was, the reinvention was a task neither long nor difficult for him. One of his first instruments magnified but three times; still it made a great sensation in Venice where he exhibited the little tube to the authorities of that city, in which he first invented it.

Galileo's telescope was of the simplest type, with but two lenses; the one a double convex lens with which an image of the distant object is formed, the other a double concave lens, much smaller which was the eye-lens for examining the image. It is this simple form of Galilean telescope that is still used in opera glasses and field glasses, because of the shorter tube necessary.

Galileo carried on the construction of telescopes, all the time improving their quality and enlarging their power until he built one that magnified thirty times. What the diameter of the object glass was we do not know, perhaps two inches or possibly a little more. Glass of a quality good enough to make a telescope of cannot have been abundant or even obtainable except with great difficulty in those early days.

Other discoveries by this first of celestial observers were the spots on the sun, the larger mountains of the moon, the separate stars of which the Milky Way is composed, and, greatest wonder of all, the anomalous "handles" (*ansæ*, he called them) of Saturn, which we now know as the planet's ring, the most wonderful of all the bodies in the sky.

Since Galileo's time, only three centuries past, the progress in size and improvement in quality of the telescope have been marvelous. And this advance would not have been possible except for, first, the discoveries still kept in large part secret by the makers of optical glass which have enabled them to make disks of the largest size; second, the consummate skill of modern opticians in fashioning these disks into perfect lenses; and third, the progress in the mechanical arts and engineering, by which telescope tubes of many tons' weight are mounted or poised so delicately that the thrust of a finger readily swerves them from one point of the heavens to another.

The 100-Inch Hooker Telescope, Largest Reflector in the World, on Mt. Wilson.

(Photo, Mt. Wilson Solar Observatory.)

The Largest Refractor, the 40-Inch Telescope at Yerkes Observatory. Dome 90 Ft. in Diameter.

(Photo, Yerkes Observatory.)

The 150-ft. Tower at the Mt. Wilson Solar Observatory. At the left is a diagram of tower, telescope and pit. At the upper right is an exterior view of the tower; below a view looking down into the pit, 75 ft. deep.

(Photo, Mt. Wilson Solar Observatory.)

As the telescope is the most important of all astronomical instruments, it is necessary to understand its construction and adjustment and how the astronomer uses it. Telescopes are optical instruments, and nothing but optical parts would be requisite in making them, if only the optical conditions of their perfect working could be obtained without other mechanical accessories.

In original principle, all telescopes are as simple as Galileo's; first, an object glass to form the image of the distant object; second the eyepiece usually made of two lenses, but really a microscope, to magnify that image, and working in the same way that any microscope magnifies an object close at hand; and third, a tube to hold all the necessary lenses in the true relative positions.

The focal lengths of object glass and eyepiece will determine just what distance apart the lenses must be in order to give perfect vision. But it is quite as important that the axes of all the lenses be adjusted into one and the same straight line, and then held there rigidly and permanently. Otherwise vision with the telescope will be very imperfect and wholly unsatisfactory. The distance from the objective, or object glass to its focal point is called its focal length; and if we divide this by the focal length of the eyepiece, we shall have the magnifying power of the telescope. The eyepiece will usually be made of two lenses, or more, and we use its focal length considered as a single lens, in getting the magnifying power. A telescope will generally have many eyepieces of different focal lengths, so that it will have a corresponding range of magnifying powers. The lowest magnifying power will be not less than four or five diameters for each inch of aperture of the objective; otherwise the eye will fail to receive all the light which falls upon the glass. A 4-inch telescope will therefore have no eyepiece with a lower magnifying power than about 20 diameters. The highest magnifying power advantageous for a glass of this size will be about 250 to 300, the working rule being about 70 diameters to each inch of aperture, although the theoretical limit is regarded as 100.

The reason for a variety of eyepieces with different magnifying powers soon becomes apparent on using the telescope. Comets and nebulæ call for very low powers, while double stars and the planetary surfaces require the higher powers, provided the state of the atmosphere at the moment will allow it. If there is much quivering and unsteadiness, nothing is gained by trying the higher powers, because all the waves of unsteadiness are magnified also in the same proportion, and sharpness of vision, or fine definition, or "good seeing," as it is called, becomes impossible. The vibrations and tremors of the atmosphere are the greatest of all obstacles to astronomical observation, and the search is always in order for regions of the world, in deserts or on high mountains, where the quietest atmosphere is to be found.

Quite another power of the telescope is dependent on its objective solely: its light-gathering power. Light by which we see a star or planet is admitted to the retina of the eye through an adjustable aperture called the pupil. In the dark or at night, the pupil expands to an average diameter of one-fourth of an inch. But the object-glass of a telescope, by focusing the rays from a star, pours into the eye, almost as a funnel acts with water, all the light which falls on its larger surface. And as geometry has settled it for us that areas of surfaces are proportioned to the squares of their diameters, a two-inch object glass focuses upon the retina of the eye 64 times as much light as the unassisted eye would receive. And the great

40-inch objective of the Yerkes telescope would, theoretically, yield 25,600 times as much light as the eye alone. But there would be a noticeable percentage of this lost through absorption by the glasses of the telescope and scattering by their surfaces.

The first makers of telescopes soon encountered a most discouraging difficulty, because it seemed to them absolutely insuperable. This is known as chromatic aberration, or the scattering of light in a telescope due simply to its color or wave length. When light passes through a prism, red is refracted the least and violet the most. Through a lens it is the same, because a lens may be regarded as an indefinite system of prisms. The image of a star or planet, then, formed by a single lens cannot be optically perfect; instead it will be a confused intermingling of images of various colors. With low powers this will not be very troublesome, but great indistinctness results from the use of high magnifying powers.

The early makers and users of telescopes in the latter part of the seventeenth century found that the troublesome effects of chromatic aberration could be much reduced by increasing the focal length of the objective. This led to what we term engineering difficulties of a very serious nature, because the tubes of great length were very awkward in pointing toward celestial objects, especially near the zenith, where the air is quietest. And it was next to impossible to hold an object steadily in the field, even after all the troubles of getting it there had been successfully overcome.

Bianchini and Cassini, Hevelius and Huygens were among the active observers of that epoch who built telescopes of extraordinary length, a hundred feet and upward. One tube is said to have been built 600 feet in length, but quite certainly it could never have been used. So-called aerial telescopes were also constructed, in which the objective was mounted on top of a tower or a pole, and the eyepiece moved along near the ground. But it is difficult to see how anything but fleeting glimpses of the heavenly bodies could have been obtained with such contrivances, even if the lenses had been perfect. Newton indeed, who was expert in optics, gave up the problem of improving the refracting telescope, and turned his energies toward the reflector.

In 1733, half a century after Newton and a century and a quarter after Galileo, Chester More Hall, an Englishman, found by experiment that chromatic aberration could be nearly eliminated by making the objective of two lenses instead of one, and the same invention was made independently by Dollond, an English optician, who took out letters patent about 1760. So the size of telescopes seemed to be limited only by the skill of the glassmaker and the size of disks that he might find it practicable to produce.

What Hall and Dollond did was to make the outer or crown lens of the objective as before, and place behind it a plano-concave lens of dense flint glass. This had the effect of neutralizing the chromatic effect, or color aberration, while at the same time only part of the refractive effect of the crown lens was destroyed. This ingenious but costly combination prepared the way for the great refracting telescopes of the present day, because it solved, or seemed to solve, the important problem of getting the necessary refraction of light rays without harmful dispersion or decomposition of them.

Through the 18th century and the first years of the 19th many telescopes of a size very great for that day were built, and their success seemed complete. With large increase in the size of the disks, however, a new trouble arose, quite inherent in the glass itself. The two kinds of glass, flint and crown, do not decompose white light with uniformity, so that when the so-called achromatic objective was composed of flint and crown, there was an effect known as irrationality of dispersion, or secondary spectrum, which produced a very troublesome residuum of blue light surrounding the images of bright objects. This is the most serious defect of all the great refractors of the day, and effectively it limits their size to about 60 inches of aperture, with present types of flint and crown. It is expected by present experimenters, however, that further improvements in optical glass will do much to extend this limit; so that a refracting telescope of much greater size than any now in existence will be practicable.

Improvements in mounting telescopes, too, are still possible. Within recent years, Hartness, of Springfield, Vermont, has erected a new and ingenious type of turret telescope which protects the observer from wind and cold while his instrument is outside. It affords exceptional facilities for rapid and convenient observing, as for variable stars, and is adaptable to both refractors and reflectors.

The captivating study of the heavens can of course be begun with the naked eye alone, but very moderate optical assistance is a great help and stimulates. An opera-glass affords such assistance; a field-glass does still better, and best of all, for certain purposes, is a modern prism-binocular.

CHAPTER XIX

REFLECTORS—MIRROR TELESCOPES

Cherished with the utmost care in the rooms of the Royal Society of London is a world-famous telescope, a diminutive reflector made by the hands of Sir Isaac Newton. We have already mentioned his connection with the refractor; and how he abandoned that type of telescope in favor of the reflecting mirror, or reflector in which the obstacles to great size appeared to be purely mechanical. By many, indeed, Newton is regarded as the inventor of the reflector.

By the principles of optics, all the rays from a star that strike a concave mirror will be reflected to the geometric focal point, provided a section of that mirror is a parabola. Such a mirror is called a speculum, and is an alloy of tin, copper, and bismuth. Its surface takes a very high polish, reflecting when newly polished nearly 90 per cent of the light that falls upon it.

But the focus where the eyepiece must be used is in front of the mirror, and if the eye were placed there, the observer's head would intercept all or much of the light that would otherwise reach the mirror. Gregory, probably the real inventor of the reflector, was the first to dodge this difficulty by perforating the mirror at the center and applying the eyepiece there, at the back of the speculum; but it was necessary to first send the rays to that point by reflection from a second or smaller mirror, in the optical axis of the speculum. This reflects the rays backward down the tube to the eyepiece, or spectroscope, or camera.

Another English optician, Cassegrain, improved on this design somewhat by placing the secondary mirror inside the focus of the speculum, or nearer to it, so that the tube is shorter. This form is preferable for many kinds of astronomical work, especially photography. Herschel sought to do away with the secondary reflector entirely and save the loss of light by tilting the speculum slightly, so as to throw the image at one side of the tube; but this modification introduces bad definition of the image and has never been much used.

A better plan is that of Newton, who placed a small plane speculum at an angle of 45 degrees in the optical axis where the secondary mirror of the Gregory-Cassegrainian type is placed. The rays are then received by the eyepiece at the side of the upper end of the tube, the observer looking in at right angles to the axis. And a modern improvement first used by Draper is a small rectangular prism in place of the little plane speculum, effecting a saving of five to ten per cent of the light.

It is not easy to say which type of telescope, the refractor or the reflector, is the more famous. Nor which is the better or more useful, or the more likely to lead in the astronomy of the future. When the successors of Dollond had carried the achromatic refractor to the limit enforced by the size of the glass disks they were able to secure, they found these instruments not so great an improvement after all. The single-lens telescopes of great focal length were nearly as good optically, though much more awkward to handle. But the quality of the glass obtainable in that day appeared to set an arbitrary limit to that great amplification of size and power which progress in observational astronomy demanded.

Then came the elder Herschel, best known and perhaps the greatest of all astronomers. At Bath, England, music was his profession, especially the organ. But he was dissatisfied with his little Gregorian reflector, and being a very clever mechanician he set out to build a reflector for himself. It is said that he cast and polished nearly 200 mirrors, in the course of experiments on the most highly reflective type of alloys, and the sort of mechanism that would enable him to give them the highest polish. In all his work he was ably and enthusiastically aided by his sister, Caroline Herschel, most famous of all women astronomers.

Upward in size of his mirrors he advanced, till he had a speculum of two feet diameter with a tube 20 feet long. Twelve to fifteen years had elapsed when in 1781, while testing one of these reflectors on stars in the constellation Gemini, he made the first discovery of a planet since the invention of the telescope—the great planet now known as Uranus.

Under the patronage of King George, he advanced to telescopes of still greater size, his largest being no less than forty feet in length, with a speculum of four feet in diameter. Two new satellites of Saturn were discovered with this giant reflector, which was dismantled by Sir John Herschel with appropriate ceremonies, including the singing of an ode by the Herschel family assembled inside of the tube, on New Year's Eve, 1839-40.

We have record of but few attempts to improve the size and definition of great reflectors by the continental astronomers during this era. In England and Ireland, however, great progress was made. About 1860 Lassell built a two-foot reflector, with which he discovered two new satellites of Uranus, and which he subsequently set up in the island of Malta. Ten years later Thomas Grubb and Son of Dublin constructed a four-foot reflector, now at the Observatory in Melbourne, Australia. Calver in conjunction with Common of Ealing, London, about 1880-95 built several large reflectors, the largest of five feet diameter, now owned by Harvard College Observatory; and, rather earlier, Martin of Paris completed a four-foot reflector.

The mirrors of these latter instruments were not made of speculum metal, but of solid glass, which must be very thick (one-seventh their diameter) in order to prevent flexure or bending by their own weight. So sensitive is the optical surface to distortion that unless a complicated series of levers and counterpoises is supplied, to support the under surface of the mirror, the perfection of its optical figure disappears when the telescope is directed to objects at different altitudes in the sky. The upper or outer surface of the glass is the one which receives the optical polish on a heavy coat of silver chemically deposited on the polished glass after its figure has been tested and found satisfactory.

But far and away the most famous reflecting telescope of all is the "Leviathan" of Lord Rosse, built at Birr Castle, Parsonstown, Ireland, about the middle of the last century. His Lordship made many ingenious improvements in grinding the mirror, which was of speculum metal, six feet in diameter and weighed seven tons. It was ground to a focal length of fifty-four feet and mounted between heavy walls of masonry, so that the motion of the great tube was restricted to a few degrees on both sides of the meridian. The huge mechanism was very cumbersome in operation, and photography was not available in those days; nevertheless Lord Rosse's telescope made the epochal discovery of the spiral nebulæ, which no other telescope of that day could have done.

In America the reflector has always kept at least even pace with the refractor. As early as 1830, Mason and Smith, two students at Yale College, enthused by Denison Olmsted, built a 12-inch speculum with which they made unsurpassed observations of the nebulæ. Dr. Henry Draper, returning from a visit to Lord Rosse, began about 1865 the construction of two silver-on-glass reflectors, one of 15 inches diameter, the other of 28 inches, with which he did important work for many years in photography and spectroscopy, and his mirrors are now the property of Harvard College Observatory. Alvan Clark and Sons have in later years built a 40-inch mirror for the Lowell Observatory in Arizona, and very recently a 6-foot silver-on-glass mirror has been set up in the Dominion of Canada Astrophysical Observatory at Victoria, British Columbia, where it is doing excellent work in the hands of Plaskett, its designer.

The huge glass disk for the reflector weighs two tons, and it must be cast so that there are no internal strains; otherwise it is liable to burst in fragments in the process of grinding. It should be free from air-bubbles, too; so the glass is cast in one melting, if possible. This disk was made by the St. Gobain Plate Glass Company, whose works have been ruthlessly destroyed by the enemy during the war; but fortunately the great disk had been shipped from Antwerp only a week before declaration of hostilities.

Brashear of Allegheny was intrusted with the optical parts, which occupied many months of critical work. The finished mirror is 73 inches in diameter, its focal length is 30 feet, and its thickness 12 inches. A central hole 10 inches in diameter makes possible its use as a Gregorian or Cassegrainian type, as well as Newtonian. The mechanical parts of this great telescope are by Warner and Swasey of Cleveland, after the well-known equatorial mounting of the Melbourne reflector by Grubb of Dublin. Friction of the polar and declination axes is reduced by ball bearings. The 66-foot dome has an opening 15 feet wide and extending six feet beyond the zenith. All motions of the telescope, dome shutters, and observing platform are under complete control by electric motors. Spectroscopic binaries form one of the special fields of research with this powerful instrument, and many new binaries have already been detected.

The great reflectors designed and constructed by Ritchey, formerly of Chicago and now of Pasadena, deserve especial mention. While connected with the Yerkes

Observatory he constructed a two-foot reflector for that institution, with which he had exceptional success in photography of the stars and nebulæ. Later he built a 5-foot reflector, now at the Carnegie Observatory on Mount Wilson, California, with which the spiral nebulæ and many other celestial objects have been especially well photographed. Ritchey's later years have been spent on the construction of an even greater mirror, no less than 100 inches in diameter, which was completed in 1919, and has already yielded photographic results dealt with farther on, and far surpassing anything previously obtained. Theoretically this huge mirror, if its surface were perfectly reflective so that it would transmit all the rays falling upon it, would gather 160,000 times as much light as the unaided eye alone.

Whether a 72-inch refractor, should it ever be constructed, would surpass the 100-inch reflector as an all-round engine for astronomical research, is a question that can only be fully answered by building it and trying the two instruments alongside.

Probably three-quarters of all the really great astronomical work in the past has been done by refractors. They are always ready and convenient for use, and the optical surfaces rarely require cleaning and readjustment. With increase of size, however, the secondary spectrum becomes very bothersome in the great lenses; and the larger they are, the more light is lost by absorption on account of the increasing thickness of the lenses. With the reflector on the other hand, while there is clearly a greater range of size, the reflective surface retains its high polish only a brief period, so that mere tarnish effectively reduces the aperture; and the great mirror is more or less ineffective in consequence of flexure uncompensated by the lever system that supports the back of the mirror.

Both types of telescope still have their enthusiastic devotees; and the next great reflector would doubtless be a gratifying success, if mounted in some elevated region of the world, like the Andes of northern Chile, where the air is exceptionally steady and the sky very clear a large part of the year. The highest magnifying powers suitable for work with such a telescope could then be employed, and new discoveries added as well as important work done in extension of lines already begun on the universe of stars.

On the authority of Clark, even a six-foot objective would not necessitate a combined thickness of its glasses in excess of six inches. Present disks are vastly superior to the early ones in transparency, and there is reason to expect still greater improvement. The engineering troubles incident to execution of the mechanical side of the scheme need not stand in the way; they never have, indeed the astronomer has but just begun to invoke the fertile resources of the modern engineer. Not long before his death the younger Clark who had just finished the great lenses of the 40-inch Yerkes telescope, ventured this prevision, already in part come true: "The new astronomy, as well as the old, demands more power. Problems wait for their solution, and theories to be substantiated or disproved. The horizon of science has been greatly broadened within the last few years, but out upon the borderland I see the glimmer of new lights that await for their interpretation, and the great telescopes of the future must be their interpreters."

Practically all the great telescopes of the world have in turn signalized the new accession of power by some significant astronomical discovery: to specify, one of Herschel's reflectors first revealed the planet Uranus; Lord Rosse's "Leviathan" the spiral nebulæ; the 15-inch Cambridge lens the crape, or dusky ring of Saturn; the 18½-inch Chicago refractor the companion of Sirius; the Washington 26-inch telescope the satellites of Mars; the 30-inch Pulkowa glass the nebulosities of the Pleiades; and the 36-inch Lick telescope brought to light a fifth satellite of Jupiter. At the time these discoveries were made, each of these great telescopes was the only instrument then in existence with power enough to have made the discovery possible. So we may advance to still farther accessions of power with the expectation that greater discoveries will continue to gratify our confidence.

CHAPTER XX

THE STORY OF THE SPECTROSCOPE

Sir Isaac Newton ought really to have been the inventor of the spectroscope, because he began by analyzing light in the rough with prisms, was very expert in optics, and was certainly enough of a philosopher to have laid the foundations of the science.

What Newton did was to admit sunlight into a darkened room through a small round aperture, then pass the rays through a glass prism and receive the band of color on a screen. He noticed the succession of colors correctly—violet, indigo, blue, green, yellow, orange, red; also that they were not pure colors, but overlapping bands of color. Apparently neither he nor any other experimenter for more than a century went any further, when the next essential step was taken by Wollaston about 1802 in England. He saw that by receiving the light through a narrow slit instead of a round hole, he got a purer spectrum, spectrum being the name given to the succession of colors into which the prism splits up or decomposes the original beam of white sunlight. This seemingly insignificant change, a narrow slit replacing the round hole, made Wollaston and not Newton the discoverer of the dark lines crossing the spectrum at various irregular intervals, and these singularly neglected lines meant the basis of a new and most important science.

Even Wollaston, however, passed them by, and it was Fraunhofer who in 1814-1815 first made a chart of them. Consequently they are known as Fraunhofer lines, or dark absorption lines. Sending the beam of light through a succession of prisms gives greater dispersion and increases the power of the spectroscope. The greater the dispersion the greater the number of absorption lines; and it is the number and intensity of these lines, with their accurate position throughout the range of the spectrum which becomes the basis of spectrum analysis.

The half century that saw the invention of the steam engine, photography, the railroad and the telegraph elapsed without any farther developments than mere mapping of the fundamental lines, A, B, C, D, E, F, G, H of the solar spectrum. The moon, too, was examined and its spectrum found the same, as was to be expected from sunlight simply reflected.

Sir John Herschel and other experimenters came near guessing the significance of the dark lines, but the problem of unraveling their mystery was finally solved by Bunsen and Kirchhoff who ascertained that an incandescent gas emits rays of exactly the same degree of refrangibility which it absorbs when white light is passed through it. This great discovery was at once received as the secure basis of spectrum analysis, and Kirchhoff in 1858 put in compact and comprehensive form the three following principles underlying the theory of the science:

(1) Solid and liquid bodies, also gases under high pressure, give when incandescent a continuous spectrum, that is one with a mere succession of colors, and neither bright nor dark lines;

(2) Gases under low pressure give a discontinuous spectrum, crossed by bright lines whose number and position in the spectrum differ according to the substances vaporized;

(3) When white light passes through a gas, this medium absorbs or quenches rays of identical wave-length with those composing its own bright-line spectrum.

Clearly then it makes no difference where the light originates whether it comes from sun or star. Only it must be bright enough so that we can analyze it with the spectroscope. But our analysis of sun and star could not proceed until the chemist had vaporized in the laboratory all the elements, and charted their spectra with accuracy. When this had been done, every substance became at once recognizable by the number and position of its lines, with practical certainty.

How then can we be sure of the chemical and physical composition of sun and stars? Only by detailed and critical comparison of their spectra with the laboratory spectra of elements which chemical and physical research have supplied. As in the sun, so in the stars, each of which is encircled by a gaseous absorptive layer or atmosphere, the light rays from the self-luminous inner sphere must pass through this reversing layer, which absorbs light of exactly the same wave-length as the lines that make up its own bright line spectrum. Whatever substances are here found in gaseous condition, the same will be evident by dark lines in the spectrum of sun or star, and the position of these dark lines will show, by coincidence with the position of the laboratory bright lines, all the substances that are vaporized in the atmospheres of the self-luminous bodies of the sky.

Here then originated the science of the new astronomy: the old astronomy had concerned itself mainly with positions of the heavenly bodies, *where* they are; the new astronomy deals with their chemical composition and physical constitution, and *what* they are. Between 1865 and 1875 the fundamental application of the basic principles was well advanced by the researches of Sir William Huggins in England, of Father Angelo Secchi in Rome, of Jules Janssen in Paris, and of Dr. Henry Draper in New York.

In analyzing the spectrum of the sun, many thousands of dark absorption lines are found, and their coincidences with the bright lines of terrestrial elements show that iron, for instance, is most prominently identified, with rather more than 2,000 coincidences of bright and dark lines. Calcium, too, is indicated by peculiar intensity of its lines, as well as their great number. Next in order are hydrogen, nickel and sodium. By prolonged and minute comparison of the solar spectrum with spectra of terrestrial elements, something like forty elemental substances are now known to exist in the sun. Rowland's splendid photographs of the solar spectrum have contributed most effectively. About half of these elements, though not in order of certainty, are aluminum, cadmium, calcium, carbon, chromium, cobalt, copper, hydrogen, iron, magnesium, manganese, nickel, scandium, silicon, silver, sodium, titanium, vanadium, yttrium, zinc, and zirconium. Oxygen, too, is pretty surely indicated; but certain elements abundant on earth, as nitrogen and chlorine, together with gold, mercury, phosphorus, and sulphur, are not found in the sun.

The two brilliant red stars, Aldebaran in Taurus, and Betelgeuse in Orion, were the first stars whose chemical constitution was revealed to the eye of man, and Sir William Huggins of London was the astronomer who achieved this epoch-making result. Father Secchi of the Vatican Observatory proceeded at once with the visual examination of the spectra of hundreds of the brighter stars, and he was the first to provide a classification of stellar spectra. There were four types.

Secchi's type I is characterized chiefly by the breadth and intensity of dark hydrogen lines, together with a faintness or entire absence of metallic lines. These are bluish or white stars and they are very abundant, nearly half of all the stars. Vega, Altair, and numerous other bright stars belong to this type, and especially Sirius, which gives to the type the name "Sirians."

Type II is characterized by a multitude of fine dark metallic lines, closely resembling the lines of the solar spectrum. These stars are somewhat yellowish in tinge like the sun, and from this similarity of spectra they are called "solars." Arcturus and Capella are "solars," and on the whole the solars are rather less numerous than the Sirians. Stars nearest to the solar system are mostly of this type, and, according to Kapteyn of Groningen, the absolute luminous power of first type stars exceeds that of second type stars seven-fold.

Secchi's type III is characterized by many dark bands, well defined on the side toward the blue end of the spectrum, but shading off toward the red—a "colonnaded spectrum", as Miss Clerke aptly terms it. Alpha Herculis, Antares, and Mira, together with orange and reddish stars and most of the variable stars, belong in type III.

Type IV is also characterized by dark bands, often called "flutings," similar to those of type III, but reversed as to shading, that is, well defined on the side toward the red, but fading out toward the blue. Their atmospheres contain carbon; but they are not abundant, besides being faint and nearly all blood-red in tint.

Following up the brilliant researches of Draper, who in 1872 obtained the first successful photograph of a star's spectrum, that of Vega, Pickering of Harvard supplemented Secchi's classification by Type V, a spectrum characterized by bright lines. They, too, are not abundant and are all found near the middle of the Galaxy. These are usually known as Wolf-Rayet stars, from the two Paris astronomers who first investigated their spectra. Type V stars are a class of objects seemingly apart from the rest of the stellar universe, and many of the planetary nebulæ yield the same sort of a spectrum.

The late Mrs. Anna Palmer Draper, widow of Dr. Henry Draper, established the Henry Draper Memorial at Harvard, and investigation of the photographic spectra of all the brighter stars of the entire heavens has been prosecuted on a comprehensive scale, those of the northern hemisphere at Cambridge, and of the southern at Arequipa, Peru. These researches have led to a broad reclassification of the stars into eight distinct groups, a work of exceptional magnitude begun by the late Mrs. Fleming and recently completed by Miss Annie Cannon, who classified the photographic spectra of more than 230,000 stars on the new system, as follows:—

The letters O, B, A, F, G, K, M, N represent a continuous gradation in the supposed order of stellar evolution, and farther subdivision is indicated by tenths, G5K meaning a type half way between G and K, and usually written G5 simply. B2 would indicate a type between B and A, but nearer to B than A, and so on. On this system, the spectrum of a star in the earliest stages of its evolution is made up of diffuse bright bands on a faint continuous background. As these bands become fewer and narrower, very faint absorption lines begin to appear, first the helium lines, followed by several series of hydrogen lines. On the disappearance of the bright bands, the spectrum becomes wholly absorptive bands and lines. Then comes a very great increase in intensity of the true hydrogen spectrum, with wide and much diffused lines, and few if any other lines. Then the H and K calcium lines and other lines peculiar to the sun become more and more intense. Then the hydrogen lines go through their long decline. The calcium spectrum becomes intense, and later the spectrum becomes quite like that of the sun with a great wealth of lines. Following this stage the spectrum shortens from the ultra violet, the hydrogen lines fade out still farther, and bands due to metallic compounds make their appearance, the entire spectrum finally resembling that of sun spots. To designate these types rather more categorically:—

Type O—bright bands on a faint continuous background, with five subdivisions, Oa, Ob, Oc, Od, Oe, according to the varying width and intensity of the bands.

Type B—the Orion type, or helium type, with additional lines of origin unknown as yet, but without any of the bright bands of type O.

Type A—the Sirian type, the regular Balmer series of hydrogen lines being very intense, with a few other lines not conspicuously marked.

Type F—the calcium type, hydrogen lines less strongly marked, but with the narrow calcium lines H and K very intense.

Type G—the solar type, with multitudes of metallic lines.

Type K—in some respects similar to G, but with the hydrogen lines fading out, and the metallic lines relatively more prominent.

Type M—spectrum with peculiar flutings due to titanium oxide, with subdivisions Ma and Mb, and the variable stars of long period, with a few bright hydrogen lines additional, in a separate class Md.

Type N—similar to M, in that both are pronouncedly reddish, but with characteristic flutings probably indicating carbon compounds.

The Draper classification being based on photographic spectra, and the original Secchi classification being visual, the relation of the two systems is approximately as follows:

Secchi Type I includes Draper B & A
Secchi Type II includes Draper F, G & K
Secchi Type III includes Draper M
Secchi Type IV includes Draper N

Pickering's marked success in organization and execution of this great programme was due to his adoption of the "slitless spectroscope," which made it possible to photograph stellar spectra in vast numbers on a single plate. The first observers of stellar spectra placed the spectroscope beyond the focus of the telescope with which it was used, thereby limiting the examination to but one star at a time. In the slitless spectroscope, a large prism is mounted in front of the objective (of short focus), so that the star's rays pass through it first, and then are brought to the same focus on the photographic plate, for all the stars within the field of view, sometimes many thousand in number. This arrangement provides great advantages in the comparison and classification of stellar spectra.

When spectroscopic methods were first introduced into astronomy, there was no expectation that the field of the old or so-called exact astronomy would be invaded. Physicists were sometimes jocularly greeted among astronomers as "ribbon men," and no one even dreamed that their researches were one day to advance to equal recognition with results derived from micrometer, meridian circle, and heliometer.

The first step in this direction was taken in 1868 by Sir William Huggins of London, who noticed small displacements in the lines of spectra of very bright stars. In fact the whole spectrum appeared to be shifted; in the case of Sirius it was shifted toward the red, while the whole spectrum of Arcturus was shifted by three times this amount toward the violet end of the spectrum. The reason was not difficult to assign.

As early as 1842 Doppler had enunciated the principle that when we are approaching or are approached by a body which is emitting regular vibrations, then the number of waves we receive in a second is increased, and their wave-length correspondingly diminished; and just the reverse of this occurs when the distance of the vibrating body is increasing. It is the same with light as with sound, and everyone has noticed how the pitch of a locomotive whistle suddenly rises as it passes, and falls as suddenly on retreating from us. So Huggins drew the immediate inference that the distance between the earth and Sirius was increasing at the rate of nearly twenty miles per second, while Arcturus was nearing us with a velocity of sixty miles per second.

These pioneer observations of motions in the line of sight, or radial velocities as they are now called, led directly to the acceptance of the high value of spectroscopic work as an adjunct of exact astronomy in stellar research. Nor has it been found wanting in application to a great variety of exact problems in the solar system which would have been wholly impossible to solve without it.

Foremost is the sun, of course, because of the overplus of light. Young early measured the displacement of lines in the spectra of the prominences, and found velocities sometimes exceeding 250 miles per second. Many astronomers, Dunér among them, investigated the rotation of the sun by the spectroscopic method. The sun's east limb is coming toward us, while the west is going from us; and by measuring the sum of the displacements, the rate of rotation has been calculated, not only at the sun's equator but at many solar latitudes also, both north and south. As was to be expected, these results agree well with the sun's rotation as found by the transits of sun spots in the lower latitudes where they make their appearance.

Bélopolsky has applied the same method to the rotation of the planet Venus, and Keeler, by measuring the displacement of lines in the spectrum of Saturn, on opposite sides of the ring, provided a brilliant observational proof of the physical constitution of the rings; because he showed that the inner ring traveled round more swiftly than the outer one, thus demonstrating that the ring could not be solid, but must be composed of multitudes of small particles traveling around the ball of Saturn, much as if they were satellites. Indeed, Keeler ascertained the velocity of their orbital motion and found that in each case it agreed exactly with that required by the Keplerian law.

Even the filmy corona of the sun was investigated in similar fashion by Deslandres at the total eclipse of 1893, and he found that it rotates bodily with the sun. But the complete vindication of the spectroscopic method as an adjunct of the old astronomy came with its application to measurement of the distance of the sun. The method is very interesting and was first suggested by Campbell in 1892. Spectrum-line measurements have become very accurate with the introduction of dry-plate photography, and ecliptic stars were spectrographed, toward and from which the earth is traveling by its orbital motion round the sun. By accurate measurement of these displacements, the orbital velocity of the earth is calculated; and as we know the exact length of the year, or a complete period, the length of the orbit itself in miles becomes known, and thus, by simple mensuration, the length of the radius of the orbit—which is the distance of the sun.

If we pass from sun to star, the triumph of the spectroscope has been everywhere complete and significant. As the spectroscopic survey of the stars grew toward completeness, it became evident that the swarming hosts of the stellar universe are in constant motion through space, not only athwart the line of vision as their proper motions had long disclosed, but some stars are swiftly moving toward our solar system and others as swiftly from it.

Fixed stars, strictly speaking—there are no such. All are in relative motion. Exact astronomy by discussion of the proper motions had assigned a region of the sky toward which the sun and planets are moving. Spectrography soon verified this direction not only, but gave a determination of the velocity of our motion of twelve miles per second in a direction approximately that of the constellation Lyra. From corresponding radial velocities, we draw the ready conclusion that certain groups or clusters of stars are actually connected in space and moving as related systems, as in the Pleiades and Ursa Major.

Rather more than a quarter century ago, the spectroscope came to the assistance of the telescope in helping to solve the intricate problem of stellar distribution. Kapteyn, by combining the proper motions of certain stars with their classification in the Draper catalogue of stellar spectra, drew the conclusion that, as stars having very small proper motions show a condensation toward the Galaxy, the stars composing this girdle are mostly of the Sirian type, and are at vast distances from the solar system. The proper motion of a star near to us will ordinarily be large, and, in the case of solar stars, the larger their proper motion the greater their number. So it would appear that the solar stars are aggregated round the sun himself, and this conclusion is greatly strengthened by the fact that of stars whose distances and spectral type are both ascertained, seven of the eight nearest to us are solar stars.

In 1889 the spectroscope achieved an unexpected triumph by enabling the late Professor Pickering to make the first discovery of a spectroscopic double, or binary star, a type of object now quite abundant. Unlike the visual binary systems whose periods are years in length, the spectroscopic binaries have short periods, reckoned in some cases in days, or hours even. If the orbit of a very close binary is seen edge on, the light of the two stars will coalesce twice in every revolution. Halfway between these points there are two times when the two stars will be moving, one toward the earth and the other from it. At all times the light of the star, in so far as the telescope shows it, proceeds from a single object.

Now photograph the star's spectrum at each of the four critical points above indicated: in the first pair the lines are sharply defined and single, because at conjunction the stars are simply moving athwart the line of sight, while at the intermediate points the lines are double. Doppler's principle completely accounts for this: the light from the receding companion is giving lines displaced toward the red, while the approaching companion yields lines displaced toward the violet. Mizar, the double star at the bend of the handle in the Great Dipper was the first star to yield this peculiar type of spectrum, and the period of its invisible companion is about 52 days. The relative velocity of the components is 100 miles a second, and applying Newton's law we find its mass exceeds that of the sun forty-fold. Capella has been found to be a spectroscopic binary; also the pole star. Spectroscopic binaries have relatively short periods, one of the shortest known being only 35 hours in length. It is in the constellation Scorpio. Beta Aurigæ is another whose lines double on alternate nights, giving a period of four days; and the combined mass of both stars is more than twice that of the sun. The catalogue of spectroscopic binaries is constantly enlarging; but thousands doubtless exist that can never be discovered by this method, as is evident if their orbits are perpendicular to the line of sight or nearly so. The history of the spectroscopic binaries is one of the most interesting chapters in astronomy, and affords a marvelous confirmation of the prediction of Bessel who first wrote of "the astronomy of the invisible."

Find a star's distance by the spectroscope? Impossible, everyone would have said, even a very few years ago. Now, however, the thing is done, and with increasing accuracy.

Adams of Mount Wilson has found, after protracted investigation, that the relative intensity of certain spectral lines varies according to the absolute brightness of a star; indeed, so close is the correspondence that the spectroscopic observations are employed to provide in certain cases a good determination of the absolute magnitude, and therefore of the distance. To test this relation, the spectroscopic parallaxes have been compared with the measured parallaxes in numerous instances, and an excellent agreement is shown. This new method is adding extensively to our knowledge of stellar luminosities and distances, and even the vast distances of globular clusters and spiral nebulæ are becoming known.

In fact, but few departments of the old astronomy are left which the new astronomy has not invaded, and this latest triumph of the spectroscope in determining accurately the distances of even the remotest stars is enthusiastically welcomed by advocates of the old and new astronomy alike.

CHAPTER XXI
THE STORY OF ASTRONOMICAL PHOTOGRAPHY

The most powerful ally of both telescope and spectroscope is photography. Without it the marvelous researches carried on with both these types of instrument would have been essentially impossible. Even the great telescopes of Herschel and Lord Rosse, notwithstanding their splendid record as optical instruments, might have achieved vastly more had photography been developed in their time to the point where the astronomer could have employed its wonderful capabilities as he does to-day. And, with the spectroscope, it is hardly too much to say that no investigator ever observes visually with that instrument any more: practically every spectrum is made a matter of photographic record first. The observing, or nowadays the measuring, is all done afterward.

All telescopes and cameras are alike, in that each must form or have formed within it an image by means of a lens or mirror. In the telescope the eye sees the fleeting image, in the camera the process of registering the image on a plate or film is known as photography. Daguerre first invented the process (silver film on a copper plate) in 1839. The year following it was first employed on the moon, in 1850 the first star was photographed, in 1851 the first total eclipse of the sun; all by the primitive daguerreotype process, which, notwithstanding its awkwardness and the great length of exposure required, was found to possess many advantages for astronomical work.

About the middle of the last century the wet plate process, so called because the sensitized collodion film must be kept moist during exposure, came into general use, and the astronomers of that period were not slow to avail themselves of the advantages of a more sensitive process, which in 1872, in the skillful hands of Henry Draper, produced the first spectrum of a star. In 1880 a nebula was first photographed, and in 1881 a comet.

Before this time, however, the new dry-plate process had been developed to the point where astronomers began to avail of its greater convenience and increased sensitiveness, even in spite of the coarseness of grain of the film. Forty years of dry-plate service have brought a wealth of advantages scarcely dreamed of in the beginning, and nearly every department of astronomical research has been enhanced thereby, while many entirely new photographic methods of investigation have been worked out.

Continued improvement in photographic processes has provided the possibility of pictures of fainter and fainter celestial objects, and all the larger telescopes have photographed stars and nebulæ of such exceeding faintness that the human eye, even if applied to the same instrument, would never be able to see them. This is because the eye, in ten or twelve seconds of keen watching, becomes fatigued and must be rested, whereas the action of very faint light rays is cumulative on the highly sensitive film; so that a continuous exposure of many hours' duration becomes readily visible to the eye on development. So a supersensitive dry plate will often record many thousand stars in a region where the naked eye can see but one.

Perhaps the greatest amplification of photography has taken place at the Harvard Observatory under Pickering, where a library of many hundred thousand plates has accumulated; and at Groningen, Holland, where Kapteyn has established an astronomical laboratory without instruments except such as are necessary to measure photographic plates, whenever and wherever taken. So it is possible to select the clearest of skies, all over the world, for exposure of the plates, and bring back the photographs for expert discussion.

Of course the sun was the celestial body first photographed, and its surpassing brilliance necessitates reduction of exposure to a minimum. In moments of exceptional steadiness of the atmosphere, a very high degree of magnification of the solar surface on the photographic plate is permitted, and the details in formation, development, and ending of sun spots are faithfully registered. Nevertheless, it cannot be said that photography has yet entirely replaced the eye in this work, and careful drawings of sun spots at critical stages in their life are capable of registering fine detail which the plate has so far been unable to record. Janssen of Paris took photographs of the solar photosphere so highly magnified that the granulation or willow-leaf structure of the surface was clearly visible, and its variations traceable from hour to hour.

The advantages of sun spot photography in ascertaining the sun's rotation, keeping count of the spots, and in a permanent record for measurement of position of the sun's axis and the spot zones, are obvious. In direct portrayal of the sun's corona during total eclipses, photography has offered superior advantages over visual sketching, in the form and exact location of the coronal streamers; but the extraordinary differences of intensity between the inner corona and its outlying extensions are such that halation renders a complete picture on a single plate practically impossible. The filamentous detail of the inner corona, and the faintest outlying extensions or streamers, the eye must still reveal directly.

In solar spectrum photography, research has been especially benefited; indeed, exact registry of the multitudinous lines was quite impossible without it. Photographic maps of the spectrum by Thollon, McClean and Rowland are so complete and accurate that no visual charts can approach them. Rowland's great photographic map of the solar spectrum spread out into a band about forty feet in length; and in the infra-red, Langley's spectrobolometer extended the invisible heat spectrum photographically to many times that length. At the other end of the spectrum, special photographic processes have extended the ultra-violet spectrum far beyond the ocular limit, to a point where it is abruptly cut off by absorption of the earth's atmosphere. On the same plate with certain regions of the sun's spectrum, the spectra of terrestrial metals are photographed side by side, and exact coincidences of lines show that about forty elemental substances known to terrestrial chemistry are vaporized in the sun.

A View of the 100-foot Dome in Which the Largest Telescope in the World is Housed. (Courtesy, Mt. Wilson Solar Observatory.)

Mount Chimborazo, Near the Equator. An observatory located on this mountain would make it possible to study the phenomena of northern and southern skies from the same point. (Courtesy, Pan-American Union.)

Lick Observatory, on the Summit of Mt. Hamilton, About Twenty-Five Miles S. W. of San Jose, California. It contains the famous Lick telescope, a 36-inch refractor.

Near View of the Eye-End of the Yerkes Telescope. The eyepiece is removed and its place taken by a photographic plate.

Young was the first to photograph a solar prominence in 1870, and twenty years later Deslandres of Paris and Hale of Chicago independently invented the spectroheliograph, by which the chromosphere and prominences of the sun, as well as the disk of the sun itself, are all photographed by monochromatic light on a single plate. Hale has developed this instrument almost to the limit, first at the Yerkes Observatory of the University of Chicago, and more recently at the Mount Wilson Observatory of the Carnegie Institution, where spectroheliograms of marvelous perfection are daily taken. It was with this instrument that Hale discovered the effect of an electro-magnetic field in sun spots which has revolutionized solar theories, a research impossible to conceive of without the aid of photography.

When we apply Doppler's principle, photography becomes doubly advantageous, whether we determine, as Dunér did and more recently Adams, the sun's own rotation and find it to vary in different solar latitudes, the equator going fastest; or apply the method to the sun's corona at the east and west limbs of the sun, which Deslandres in 1893 proved to be rotating bodily with the sun, because of the measured displacement of spectral lines of the corona in juxtaposition on the photographic plate.

In the solar astronomy of measurement, too, photography has been helpfully utilized, as in registering the transits of Mercury over the sun's disk, for correcting the tables of the planet's orbital motion; and most prominently in the action taken by the principal governments of the world in sending out expeditions to observe the transits of Venus in 1874 and 1882, for the purpose of determining the parallax of Venus and so the distance of the earth from the sun.

In our studies of the moon, photography has almost completely superseded ocular work during the past sixty years. Rutherfurd and Draper of New York about 1865 obtained very excellent lunar photographs with wet plates, which were unexcelled for nearly half a century. The Harvard, Lick, and Paris Observatories have published pretty complete photographic atlases of the moon, and the best negatives of these series show nearly everything that the eye can discern, except under unusual circumstances. Later lunar photography was taken up at the Yerkes Observatory, and exceptionally fine photographs on a large scale were obtained with the 40-inch refractor, using a color screen. More recently the 60-inch and 100-inch mirrors of the Mount Wilson Observatory have taken a series of photographs of the moon far surpassing everything previously done, as was to be expected from the unique combination of a tranquil mountain atmosphere with the extraordinary optical power of the instruments, and a special adaptation of photographic methods. During lunar eclipses, Pickering has made a photographic search for a possible satellite of the moon, occultations of stars by the moon have been recorded by photography, and Russell of Princeton has shown how the position of the moon among the stars can be determined by the aid of photography with a high order of precision.

The story of planetary photography is on the whole disappointing. Much has been done, but there is much that is within reach, or ought to be, that remains undone. From Mercury nothing ought perhaps to be expected. On many of the photographs of the transit of Venus, especially those taken under the writer's direction at the Lick Observatory in 1882, we have unmistakable evidence of the planet's atmosphere. Here again the wet plate process, although more clumsy, demonstrated its superiority over the dry process used by other expeditions.

In spectroscopy, Bélopolsky has sought to determine the period of rotation of Venus on her axis. At the Lowell Observatory, Douglass succeeded in photographing the faint zodiacal light, and very successful photographs of Mars were taken at this institution as early as 1905 by Slipher. Two years later these were much improved upon by the writer's expedition to the Andes of Chile, when 12,000 exposures of Mars were made, many of them showing the principal *canali*, and other prominent features of the planet's disk. At subsequent oppositions of the planet, Barnard at the Yerkes Observatory and the Mount Wilson observers have far surpassed all these photographs.

For future oppositions a more sensitive film is highly desired, in connection with instruments possessing greater light-gathering power, so permitting a briefer exposure that will be less influenced by irregularities and defects of the atmosphere. The spectrum of Mars is of course that of sunlight, very much reduced, and modified to a slight extent by its passing twice through the atmosphere of Mars. What amount of aqueous vapor that atmosphere may contain is a question that can be answered only by critical comparison of the Martian spectrum with the spectrum of the moon, and photography affords the only method by which this can be done.

Many are the ways in which photography has aided research on the asteroid group. Since 1891 more than 600 of them have been discovered by photography, and it is many times easier to find the new object on the photographic plate than to detect it in the sky as was formerly done by means of star charts. The planet by its motion during the exposure of the plate produces a trail, whereas the surrounding stars are all round dots or images. Or by moving the plate slightly during exposure, as in Metcalf's ingenious method, we may catch the planet at that point where it will give a nearly circular image, and thus be quite as easy to detect, because all the stars on the same plate will then be trails.

Photographic photometry of the asteroids has revealed marked variations in their light, due perhaps to irregularities of figure. On account of their faint light, the asteroids are especially suited, as Mars is not, to exact photography for ascertaining their parallax, and from this the sun's distance when the asteroid's distance has been found. Many asteroids have been utilized in this way, in particular Eros (433). In 1931 it approaches the earth within 13 million miles, when the photographic method will doubtless give the sun's distance with the utmost accuracy.

Photographs of Jupiter have been very successfully taken at the Yerkes and Lowell Observatories and elsewhere, but the great depth of the planet's atmosphere is highly absorptive, so that the impression is very weak in the neighborhood of the limb, if the exposure is correctly timed for the center of the disk. The striking detail of the belts, however, is excellently shown. Wood of Baltimore has obtained excellent results by monochromatic photography of Jupiter and Saturn with the 60-inch reflector on Mount Wilson. Jupiter's satellites have not been neglected photographically, and Pickering has observed hundreds of the eclipses of the satellites by a sort of cinematographic method of repeated exposures, around the time of disappearance and reappearance by eclipse. The newest outer satellites of Jupiter were all discovered by photography, and it is extremely doubtful if they would have been found otherwise.

Saturn has long been a favorite object with the astronomical photographer, and there are many fine pictures in spite of his yellowish light, relatively weak photographically. The marvelous ring system with the Cassini division, the oblateness of the ball, the occasional markings on it—all are well shown in the best photographs; but the call is for more light and a more sensitive photographic process. Pickering's ninth satellite (Phœbe) was discovered by photography, one of the faintest moons in the solar system. Like the faint outer moons of Jupiter, few existing telescopes are powerful enough to show it. Its orbit has been found from photographic observations, and its position is checked up from time to time by photography.

But the crowning achievement of spectrum photography in the Saturnian system is Keeler's application of Doppler's principle in determining the rate of orbital motion of particles in different zones of the rings, thereby establishing the Maxwellian theory of the constitution of the rings beyond the possibility of doubt. For Uranus and Neptune photography has availed but little, except to negative the existence of additional satellites of these planets, which doubtless would have been discovered by the thorough photographic search which has been made for them by W. H. Pickering without result.

As with the asteroids, so with comets: several of these bodies have been discovered by photography; none more spectacular than the Egyptian comet of May 17th, 1882, which impressed itself on the plates of the corona of that date. Withdrawal of the sun's light by total eclipse made the comet visible, and it had never been seen before, nor is it known whether it will ever return. In cometary photography, much the same difficulties are present as in photographing the corona: if the plate is exposed long enough to get the faint extensions of the tail, the fine filaments of the coma or head are obliterated by halation and overexposure.

No one has had greater success in this work than Barnard, whose photographs of comets, particularly at the Lick Observatory, are numerous and unexcelled. His photographs of the Brooks Comet of 1893 revealed rapid and violent changes in the tail, as if shattered by encounter with meteors; and the tail of Halley's comet in 1910 showed the rapid propagation of luminous waves down the tail, similar to phenomena sometimes seen in streamers of the aurora. Draper obtained the first photograph of a comet's spectrum in 1881, disclosing an identity with hydrocarbons burning in a Bunsen flame, also bands in the violet due to carbon compounds. The photographic spectra of subsequent comets have shown bright lines due to sodium and the vapor of iron and magnesium.

Even the elusive meteor has been caught by photography, first by Wolf in 1891, who was exposing a plate on stars in the Milky Way. On developing it, he found a fine, dark nearly uniform line crossing it, due to the accidental flight across the field of a meteor of varying brightness. Since then meteor trails have been repeat-edly photographed, and even the trail spectra of meteors have been registered on the Harvard plates. At Yale in 1894 Elkin employed a unique apparatus for secur-ing photographic trails of meteors: six photographic cameras mounted at different angles on a long polar axis driven by clockwork, the whole arranged so as to cover a large area of the sky where meteors were expected.

When we pass from the solar system to the stellar universe the advantages of photography and the amplification of research due to its employment as accessory in nearly every line of investigation are enormous. So extensively has photography been introduced that plates, and to a slight extent films, are now almost exclusive-ly used in securing original records. Regrettably so in case of the nebulæ, because the numerous photographs of the brighter nebulæ taken since 1880 when Draper got the first photograph of the nebula of Orion, are as a rule not comparable with each other. Differences of instruments, of plates, of exposure, and development—all have occasioned differences in portrayal of a nebula which do not exist. When we consider faithful accuracy of portrayal of the nebulæ for purposes of critical comparison from age to age, many of our nebular photographs of the past forty years, fine as they are and marvelous as they are, must fail to serve the purpose of revealing progressive changes in nebular features in the future.

Roberts and Common in England were among the first to obtain nebular pho-tographs with extraordinary detail, also the brothers Henry of Paris. As early as 1888 Roberts revealed the true nature of the great nebula in Andromeda, which had never been suspected of being spiral; and Keeler and Perrine at the Lick Ob-servatory pushed the photographic discovery of spiral nebulæ so far that their estimates fill the sky with many hundred thousands of these objects.

In the southern hemisphere the 24-inch Bruce telescope of Harvard College Observatory has obtained many very remarkable photographs of nebulæ, partic-ularly in the vicinity of Eta Carinæ. But the great reflectors of the Mount Wilson Observatory, on account of their exceptional location and extraordinary power, have surpassed all others in the photographic portrayal of these objects, especially of the spiral nebulæ which appear to show all stages in transition from nebula to star. No less remarkable are the photographs of such wonderful clusters as Omega Centauri, a perfect visual representation of which is wholly impossible. Intercom-parison of the photographs of clusters has afforded Bailey of Harvard, Shapley of Mount Wilson and others the opportunity of discovery that hundreds of the component stars are variable.

What is the longest photographic exposure ever made? At the Cape of Good Hope, under the direction of the late Sir David Gill, exposures on nebulæ were made, utilizing the best part of several nights, and totaling as high as seventeen, or even twenty-three hours. But the Mount Wilson observers have far surpassed this duration. To study the rotation and radial velocity of the central part of the nebula of Andromeda, an exposure of no less than 79 hours' total duration was made on the exceedingly faint spectrum, and even that record has since been exceeded. The eye cannot be removed from the guiding star for a moment while the exposure is in progress, and this tedious piece of work was rewarded by determining the velocity of the center of the nucleus as a motion of approach at the rate of 316 kilometers per second.

But when the stars, their magnitudes and their special peculiarities are to be investigated *en masse*, photography provides the facile means for researches that would scarcely have been dreamed of without it. The international photographic chart of the entire heavens, in progress at twenty observatories since 1887, the photographic charts of the northern heavens at Harvard and of the southern sky at Cape Town, the manifold investigations that have led up to the Harvard photometry, and the unparalleled photographic researches of the Henry Draper Memorial, enabling the spectra of many hundred thousand stars to be examined and classified—all this is but a part of the astronomical work in stellar fields that photography has rendered possible.

Then there are the stellar parallaxes, now observed for many stars at once photographically, when formerly only one star's parallax could be measured at a time and with the eye at the telescope. And photo-electric photometry, measuring smaller differences of light than any other method, and providing more accurate light-curves of the variable stars. And perhaps most remarkable of all, the radial velocity work on both stars and nebulæ, giving us the distance of whole classes of stars, discovering large numbers of spectroscopic binaries and checking up the motion of the solar system toward Lyra within a fraction of a mile per second.

All told, photography has been the most potent adjunct in astronomical research, and it is impossible to predict the future with more powerful apparatus and photographic processes of higher sensitiveness. The field of research is almost boundless, and the possibilities practically without limit.

What would Herschel have done with £100,000—and photography!

CHAPTER XXII
MOUNTAIN OBSERVATORIES

The century that has elapsed since the time of Sir William Herschel, known as the father of the new or descriptive astronomy, has witnessed all the advances of the science that have been made possible by adopting the photographic method of making the record, instead of depending upon the human eye. Only one eye can be looking at the eyepiece at a time: the photograph can be studied by a thousand eyes.

At mountain elevations telescopes are now extensively employed, and there the camera is of especial and additional value, because the photograph taken on the mountain can be brought down for the expert to study, at ease and in the comfort of a lower elevation. We shall next trace the movement that has led the astronomer to seek the summits of mountains for his observatories, and the photographer to follow him.

Not only did the genius of Newton discover the law of universal gravitation, and make the first experiments in optics essential to the invention of the spectroscope, but he was the real originator also of the modern movement for the occupation of mountain elevations for astronomical observatories. His keen mind followed a ray of light all the way from its celestial source to the eye of the observer, and analyzed the causes of indistinct and imperfect vision.

Endeavoring to improve on the telescope as Galileo and his followers had left it, he found such inherent difficulties in glass itself that he abandoned the refracting type of telescope for the reflector, to the construction of which he devoted many years. But he soon found out, what every astronomer and optician knew to their keen regret, that a telescope, no matter how perfectly the skill of the optician's hand may make it, cannot perform perfectly unless it has an optically perfect atmosphere to look through.

So Newton conceived the idea of a mountain observatory, on the summit of which, as he thought, the air would be not only cloudless, but so steady and equable that the rays of light from the heavenly bodies might reach the eye undisturbed by atmospheric tremors and quiverings which are almost always present in the lower strata of the great ocean of air that surrounds our planet.

This is the way Newton puts the question in his treatise on *Opticks*—he says: "The Air through which we look upon the Stars, is in a perpetual Tremor; as may be seen by the tremulous Motion of Shadows cast from high Towers, and by the twinkling of the fix'd stars…. The only remedy is a most serene and quiet Air, such as may perhaps be found on the tops of the highest Mountains above the grosser Clouds."

Newton's suggestion is that the *highest* mountains may afford the best conditions for tranquillity; and it is an interesting coincidence that the summits of the highest mountains, about 30,000 feet in elevation, are at about the same level where the turbulence of the atmosphere most likely ceases, according to the indications of recent meteorological research. These heights are far above any elevations permanently occupied as yet, but a good beginning has been made and results of great value have already been reached.

Curiously, investigation of mountain peaks and their suitability for this purpose was not undertaken till nearly two centuries after Newton, when Piazzi Smyth in 1856 organized his expedition to the summit of a mountain of quite moderate elevation, and published his "Teneriffe: an Astronomer's Experiment." Teneriffe is an accessible peak of about 10,000 feet, on an island of the Canaries off the African coast, where Smyth fancied that conditions of equability would exist; and on reaching the summit with his apparatus and spending a few days and nights there, he was not disappointed. Could he have reached an elevation of 13,000 feet, he would have had fully one-third of all the atmosphere in weight below him, and that the most turbulent portion of all. Nevertheless, the gain in steadiness of the atmosphere, providing "better seeing," as the astronomer's expression is, even at 10,000 feet, was most encouraging, and led to attempts on other peaks by other astronomers, a few of whom we shall mention.

Davidson, an observer of the United States Coast Survey, with a broad experience of many years in mountain observing, investigated the summit of the Sierra Nevada mountains as early as 1872, at an elevation of 7,200 feet. His especial object was to make an accurate comparison between elevated stations at different heights. He found the seeing excellent, especially on the sun; but the excessive snowfall at his station, 45 feet annually, was a condition very adverse to permanent occupation.

In the summer of 1872, Young spent several weeks at Sherman, Wyoming, at an elevation exceeding 8,300 feet. He carried with him the 9.4-inch telescope of Dartmouth College, where he was then professor, and this was the first expedition on which a large glass was used by a very skillful observer at great elevation. He found the number of good days and nights small, but the sky was exceedingly favorable when clear. Many 7th magnitude stars could be detected with the naked eye. Young's observations at Sherman were mainly spectroscopic, however, and they demonstrated the immense advantage of a high-level station, far above the dust and haze of the lower atmosphere. He pronounced the 9.4-inch glass at 8,000 feet the full equivalent of a 12-inch at sea level.

Mont Blanc of 15,000 feet elevation was another summit where the veteran Janssen of Paris maintained a station for many years; but the continental conditions of atmospheric moisture and circulation were not favorable on the whole. Janssen was mainly interested in the sun, and the daylight seeing is rarely benefited, owing to the strong upward currents of warm air set in motion by the sun itself.

Mountains in the beautiful climate of California were among the earliest investigated, and when in 1874 the trustees of Mr. James Lick's estate were charged with equipping an observatory with the most powerful telescope in existence, they wisely located on the summit of Mount Hamilton. It is 4,300 feet above sea level, and Burnham and other astronomers made critical tests of the steadiness of vision there by observing double stars, which afford perhaps the best means of comparing the optical quality of the atmosphere of one region with another. The writer was fortunate in having charge of the observations of the transit of Venus in 1882 on the mountain, when the Observatory was in process of construction, and the quality of the photographs obtained on that occasion demonstrated anew the excellence of the site. Particularly at night, for about nine months of the year, the seeing is exceptionally good, especially when fog banks rolling in from the Pacific, cover the valleys below like a blanket, preventing harmful radiation from the soil below.

The great telescope mounted in 1888, a 36-inch refractor by Alvan Clark, has fulfilled every expectation of its projectors, and justified the selection of the site in every particular. The elevation, although moderate, is still high enough to secure very marked advantage in clearness and steadiness of the air, and at the same time not so high that the health and activities of the observers are appreciably affected by the thinner air of the summit. This telescope is known the world over for the monumental contributions to science made by the able astronomers who have worked with it: among them Barnard who discovered the fifth satellite of Jupiter in 1892; Burnham, Hussey, and Aitken, who have discovered and measured thousands of close double stars; Keeler, who spent many faithful years on the summit; and Campbell, the present director, whose spectroscopic researches on stellar movements have added greatly to our knowledge of the structure of the universe. Among the many lines of research now in progress at the Lick Observatory and in the D. O. Mills Observatory at Santiago, Chile, are the discoveries of stars whose velocities in space are not constant, but variable with the spectral type of the star. Mr. Lick's bequest for the Observatory was about $700,000. So ably has this scientific trust been administered that he might well have endowed it with his entire estate, exceeding $4,000,000.

Another California mountain that was early investigated is Mount Whitney. Its summit elevation is nearly 15,000 feet, and in 1881 Langley made its ascent for the purpose of measuring the solar constant. He found conditions much more favorable than on Mount Etna, Sicily—elevation about 10,000 feet—which he had visited the year before. But the height of Mount Whitney was such as to occasion him much inconvenience from mountain sickness, an ailment which is most distressing and due partly to lack of oxygen and partly to mere diminution of mechanical pressure. Mount Whitney was also visited many years after by Campbell for investigating the spectrum of Mars in comparison with that of the moon. Langley found on Mount Whitney an excellent station lower down, at about 12,000 feet elevation; and by equipping the two stations with like apparatus for measuring the solar heat, he obtained very important data on the selective absorption of the atmosphere.

Returning from the transit of Venus in 1882, Copeland of Edinburgh visited several sites in the Andes of Peru, ascending on the railway from Mollendo. Vincocaya was one of the highest, something over 14,000 feet elevation. His report was most enthusiastic, not only as to clearness and transparency of the atmosphere, but also as to its steadiness, which for planetary and double star observations is almost as important. Copeland's investigation of this region of the Andes has led many other astronomers to make critical tests in the same general region. Climatic conditions are particularly favorable, and the sites for high-level research are among the best known, the atmosphere being not only clear a large part of the year, but in certain favored spots exceedingly steady.

In 1887 the writer ascended the summit of Fujiyama, Japan, 12,400 feet elevation. The early September conditions as to steadiness of atmosphere were extraordinarily fine, but the mountain is covered by cloud many months in each year. There is a saddle on the inside of the crater that would form an ideal location for a high-level observatory. This expedition was undertaken at the request of the late Professor Pickering, director of Harvard College Observatory, which had recently received a bequest from Uriah A. Boyden, amounting to nearly a quarter of a million dollars, to "establish and maintain, in conjunction with others, an astronomical observatory on some mountain peak."

Great elevations were systematically investigated in Colorado and California, the Chilean desert of Atacama was visited, and a temporary station established at Chosica, Peru, elevation about 5,000 feet. Atmospheric conditions becoming unfavorable, a permanent station was established in 1891 at Arequipa, Peru, elevation 8,000 feet, which has been maintained as an annex to the Harvard Observatory ever since. The cloud conditions have been on the whole less favorable than was expected, but the steadiness of the air has been very satisfactory. In addition to planetary researches conducted there in the earlier years by W. H. Pickering, many large programs of stellar research have been executed, especially relating to the magnitudes and spectra of the stars. In conjunction with the home observatory in the northern hemisphere, this afforded a vast advantage in embracing all the stars of the entire heavens, on a scale not attempted elsewhere. The Bruce photographic telescope of 24-inch aperture has been employed for many years at Arequipa, and with it the plates were taken which enabled Pickering to discover the ninth satellite of Saturn (Phœbe), and the splendid photographs of southern globular clusters in which Bailey has found numerous variable stars of very short periods—very faint objects, but none the less interesting, and of much significance in modern study of the evolution and structure of the stellar universe. The crowning research of the observatory is the Henry Draper catalogue of stellar spectra, now in process of publication, which is of the first order of importance in statistical studies of stellar distribution with reference to spectral type, and in studying the relation of parallax and distance, proper motion, radial velocity and its variation to the spectral characteristics of the stars.

Perrine of Cordova is now establishing on Sierra Chica about twenty-five miles southwest of Cordova, a great reflecting telescope comparable in size with the instruments of the northern hemisphere, for investigation of the southern nebulæ and clusters, and motions of the stars. The elevation of this new Argentine observatory will be 4,000 feet above sea level.

Another observatory at mountain elevation and in a highly favorable climate is the Lowell Observatory, located at about 7,000 feet elevation at Flagstaff, Arizona. Many localities were visited and the atmosphere tested especially for steadiness, an optical quality very essential for research on the planetary surfaces. Mexico was one of these stations, but local air currents and changes of temperature there were such that good seeing was far from prevalent, as had been expected. At Flagstaff, on the other hand, conditions have been pretty uniformly good, and an enormous amount of work on the planet Mars has been accumulated and published. The first successful photographs of this planet were taken there in 1905, and Jupiter, Saturn, the zodiacal light and many other test objects have been photographed, which demonstrates the excellence of the site for astronomical research. Within recent years spectrum research by Slipher, especially on the nebulæ, has been added to the program, and the rotation and radial velocities of many nebulæ have been determined.

On Mount Wilson, near Pasadena, California, at an elevation of nearly 6,000 feet, is the Carnegie Solar Observatory, founded and equipped under the direction of Professor George E. Hale, as a department of the Carnegie Institution of Washington, of which Dr. John Campbell Merriam is President. The climatology of the region was carefully investigated and tests of the seeing made by Hussey and others. Although equipped primarily for study of the sun, the program of the observatory has been widely amplified to include the stars and nebulæ. The instrumental equipment is unique in many respects. To avoid the harmful effect of unsteadiness of air strata close to the ground a tower 150 feet high was erected, with a dome surmounting it and covering a cœlostat with mirror for reflecting the sun's rays vertically downward. Underneath the tower a dry well was excavated to a depth equal to ½ the height of the tower above it. In the subterranean chamber is the spectroheliograph of exceptional size and power. The sun's original image is nearly 17 inches in diameter on the plate, and the solar chromosphere and prominences, together with the photosphere and faculæ, are all recorded by monochromatic light.

Connected with the observatory on Mount Wilson are the laboratories, offices and instrument shops in Pasadena, 16 miles distant, where the remarkable apparatus for use on the mountain is constructed. A reflecting telescope with silver-on-glass mirror 60 inches in diameter was first built by Ritchey and thoroughly tested by stellar photographs. Also the northern spiral nebulæ were photographed, exhibiting an extraordinary wealth of detail in apparent star formation. The success of this instrument paved the way for one similar in design, but with a mirror 100 inches in diameter, provided by gift of the late John D. Hooker of Los Angeles. The telescope was completed in 1919. Notwithstanding its huge size and enormous weight, the mounting is very successful, as well as the mirror. Mercurial bearings counterbalance the weight of the polar axis in large part. This great telescope, by far the largest and most powerful ever constructed, is now employed on a program of research in which its vast light-gathering power will be utilized to the full. Under the skillful management of Hale and his enthusiastic and capable colleagues, the confines of the stellar heavens will be enormously extended, and secrets of evolution of the universe and of its structure no doubt revealed.

In all the mountain stations hitherto established, as the Lick Observatory at 4,000 feet, the Mount Wilson Observatory at 6,000 feet, the Lowell Observatory at 7,000 feet, the Harvard Observatory at 8,000 feet; and Teneriffe and Etna at 10,000, Fujiyama at 12,000, Pike's Peak at 14,000, Mont Blanc and Mount Whitney at 15,000, the researches that have been carried on have fully demonstrated the vast advantage of increased elevation in localities where climatological conditions as well as elevation are favorable. Nevertheless, only one-half of the extreme altitude contemplated by Sir Isaac Newton has yet been attained.

Can the greater heights be reached and permanently occupied? Geographically and astronomically the most favorably located mountain for a great observatory is Mount Chimborazo in Ecuador. Its elevation is 22,000 feet, and it was ascended by Edward Whymper in 1880. Situated very nearly on the earth's equator, almost the entire sidereal heavens are visible from this single station, and all the planets are favored by circumzenith conditions when passing the meridian. No other mountain in the world approaches Chimborazo in this respect. But the summit is perpetually snow-capped, exceedingly inaccessible, and the defect of barometric pressure would make life impossible up there in the open.

Only one method of occupation appears to be feasible. The permanent snow line is at about 16,000 feet, where excellent water power is available. By tunneling into the mountain at this point, and diagonally upward to the summit, permanent occupation could be accomplished, at a cost not to exceed one million dollars.

The rooms of the summit observatory would need to be built as steel caissons, and supplied with compressed air at sea-level tension. The practicability of this plan was demonstrated by the writer in September, 1907, at Cerro de Pasco, Peru. A steel caisson was carried up to an elevation exceeding 14,000 feet. Patients suffering acutely with mountain sickness were placed inside this caisson, and on restoring the atmospheric pressure within it artificially all unfavorable symptoms— headache, high respiration and accelerated pulse—disappeared. There was every indication that if persons liable to this uncomfortable complaint were brought up to this elevation, or indeed any attainable elevation, under unreduced pressure, the symptoms of mountain sickness would be unknown. Comfortable occupation of the highest mountain summits was thereby assured.

The working of astronomical instruments from within air-tight compartments does not present any insurmountable difficulties, either mechanical or physical. Since the time these experiments were made, the Guayaquil-Quito railway has been constructed over a saddle of Chimborazo, at an elevation of 12,000 feet; and only six miles of railway would need to be built from this station to the point where the tunnel would enter the mountain.

Only by the execution of some such plan as this can astronomers hope to overcome the baleful effects of an ever mobile atmosphere, and secure the advantages contemplated by Sir Isaac Newton in that tranquillity of atmosphere, which he conceived as perpetually surrounding the summits of the highest mountains.

In Russell's theory of the progressive development of the stars, from the giant class to the dwarf, an element of verification from observation is lacking, because hitherto no certain method of measuring the very minute angular diameters of the stars has been successfully applied. The apparent surface brightness corresponding to each spectral type is pretty well known, and by dividing it into the total apparent brightness, we have the angular area subtended by the star, quite independent of the star's distance. This makes it easy to estimate the angular diameter of a star, and Betelgeuse is the one which has the greatest angular diameter of all whose distances we know. Antares is next in order of angular diameter, 0".043, Aldebaran 0".022, Arcturus 0".020, Pollux 0".013, and Sirius only 0".007.

Can these theoretical estimates be verified by observation? Clearly it is of the utmost importance and the exceedingly difficult inquiry has been undertaken with the 100-inch reflector on Mount Wilson, employing the method of the interferometer developed by Michelson and described later on, an instrument undoubtedly capable of measuring much smaller angles than can be measured by any other known method. Unquestionably the interference of atmospheric waves, or in other words what astronomers call "poor seeing," will ultimately set the limit to what can be accomplished. "But even if," says Eddington, "we have to send special expeditions to the top of one of the highest mountains in the world, the attack on this far-reaching problem must not be allowed to languish."

CHAPTER XXIII
THE PROGRAM OF A GREAT OBSERVATORY

The Mount Wilson Observatory has now been in operation about fifteen years. The novelty in construction of its instruments, the investigations undertaken with them and the discoveries made, the interpretation of celestial phenomena by laboratory experiment, and the recent addition to its equipment of a telescope 100 inches in diameter, surpassing all others in power, directs especial attention to the extensive activities of this institution, whose budget now exceeds a million dollars annually. Results are only achieved by a carefully elaborated program, such as the following, for which the reader is mainly indebted to Dr. Hale, the director of the observatory, who gives a very clear idea of the trend of present-day research on the magnetic nature of the sun, and the structure and evolution of the sidereal universe.

The purpose of the observatory, as defined at its inception, was to undertake a general study of stellar evolution, laying especial emphasis upon the study of the sun, considered as a typical star; physical researches on stars and nebulæ; and the interpretation of solar and stellar phenomena by laboratory experiments. Recognizing that the development of new instruments and methods afforded the most promising means of progress, well-equipped machine shops and optical shops were provided with this end in view.

The original program of the observatory has been much modified and extended by the independent and striking discovery by Campbell and Kapteyn of an important relationship between stellar speed and spectral type; the demonstration by Hertzsprung and Russell of the existence of giant and dwarf stars; the successful application of the 60-inch reflector by Van Maanen to the measurement of minute parallaxes of stars and nebulæ; the important developments of Shapley's investigation of globular star clusters; the possibilities of research resulting from Seares's studies in stellar photometry; and the remarkable means of attack developed by Adams through the method of spectroscopic parallaxes.

By this method the absolute magnitude, and hence the distance of a star is accurately determined from estimates of the relative intensities of certain lines in stellar spectra. Attention was first directed toward lines of this character in 1906, when it was inferred that the weakening of some lines in the spectra of sun spots and the strengthening of others was the result of reduced temperature of the spot vapors. On testing this hypothesis by laboratory experiments, it was fully verified.

Subsequently Adams, who had thus become familiar with these lines and their variability, studied them extensively in the spectra of other stars. In this way was discovered the dependence of their relative intensities on the star's absolute magnitude, so providing the powerful method of spectroscopic parallaxes.

This method, giving the absolute magnitude as well as the distance of every star (excepting those of the earliest type) whose spectrum is photographed, is no less important from the evolutional than from the structural point of view.

Investigations in solar physics which formerly held chief place in the research program have developed along unexpected lines. It could not be foreseen at the outset that solar magnetic phenomena might become a subject of inquiry, demanding special instrumental facilities, and throwing light on the complex question of the nature of the sun spots and other solar problems of long standing. It is obvious that these researches, together with those on the solar rotation and the motions of the solar atmosphere, developed by Adams and St. John, must be carried to their logical conclusion, if they are to be utilized to the fullest in interpreting stellar and nebular phenomena.

The discovery of solar magnetism, like many other Mount Wilson results, was the direct outcome of a long series of instrumental developments. The progressive improvement and advance in size of the tools of research was absolutely necessary. Hale's first spectroheliograph at Kenwood in 1890 was attached to a 12-inch refractor, and the solar image was but two inches in diameter. It was soon found that a larger solar image was essential, and a spectrograph of much greater linear dispersion; in fact, the spectrograph must be made the prime element in the combination, and the telescope so designed as to serve as a necessary auxiliary.

Accordingly, successive steps have led through spectrographs of 18 and 30 feet dimension to a vertical spectrograph 75 feet in focal length. The telescope is the 150 feet tower telescope, giving a solar image of 16.5 inches in diameter. Its spectrograph is massive in construction, and by extending deep into the earth, it enjoys the stability and constancy of temperature required for the most exacting work.

Another direct outgrowth of the work of sun-spot spectra is a study of the spectra of red stars, where the chemistry of these coolest regions of the sun is partially duplicated. The combination of titanium and oxygen, and the significant changes of line intensity already observed in both instances, and also in the electric furnace at reduced temperatures, give indication of what may be expected to result from an attack on the spectra of the red stars with more powerful instrumental means, which is now provided by the 100-inch telescope and its large stellar spectrograph.

Other elements in the design of the 100-inch Hooker telescope have the same general object in view—that of developing and applying in astronomical practice the effective research methods suggested by recent advances in physics. Fresh possibilities of progress are constantly arising, and these are utilized as rapidly as circumstances permit.

The policy of undertaking the interpretations of celestial phenomena by laboratory experiments, an important element in the initial organization of Mount Wilson, has certainly been justified by its results. Indeed, the development of many of the chief solar investigations would have been impossible without the aid of special laboratory studies, going hand in hand with the astronomical observations. So indispensable are such researches, and so great is the promise of their extension, that the time has now come for advancing the laboratory work from an accessory feature to full equality with the major factors in the work of the observatory. Accordingly a new instrument now under installation is an extremely powerful electro-magnet, designed by Anderson for the extension of researches on the Zeeman effect, and for other related investigations. Within the large and uniform field of this magnet, which is built in the form of a solenoid, a special electric furnace, designed for this purpose by King, is used for the study of the inverse Zeeman effect at various angles with the lines of force. This will provide the means of interpreting certain remarkable anomalies in the magnetic phenomena of sun spots.

The 100-inch telescope is now in regular use. All the tests so far applied show that it greatly surpasses the 60-inch telescope in every class of work. For many months most of the observations and photographs have been made with the Cassegrain combination of mirrors, giving an equivalent focal length of 134 feet and involving three reflections of light. The 100-inch telescope is found to give nearly 2.8 times as much light as the 60-inch telescope, and therefore extends the scope of the instrument to all the stars an entire magnitude fainter. This is a very important gain for research on the faint globular clusters, as well as the small and faint spiral and planetary nebulæ, providing a much larger scale for these objects and sufficient light at the same time. Photographs of the moon and many other less critical tests have been made with very satisfactory results. Those of the moon appear to be decidedly superior in definition to any previously taken with other instruments.

Another investigation is of great importance in the light of recent advances in theoretical dynamics. Darwin, in his fundamental researches on the dynamics of rotating masses, dealt with incompressible matter, which assumes the well-known pear-shaped figure, and may ultimately separate into two bodies. Roche on the other hand discussed the evolution of a highly compressible mass, which finally acquires a lens-shaped form and ejects matter at its periphery. Both of these are extreme cases. Jeans has recently dealt with intermediate cases, such as are actually encountered in stars and nebulæ. He finds that when the density is less than about one-fourth that of water, a lens-shaped figure will be produced with sharp edges, as depicted by Roche. Matter thrown off at opposite points on the periphery, under the influence of small tidal forces from neighboring masses, may take the form of two symmetric filaments, though it is not yet entirely clear how these may attain the characteristic configuration of spiral nebulæ. The preliminary results of Van Maanen indicate motion outward along the arms, in harmony with Jeans's views.

Jeans further discusses the evolution of the arms, which will break up into nuclei (of the order of mass of the sun) if they are sufficiently massive, but will diffuse away if their gravitational attraction is small. The mass of our solar system is apparently not great enough, according to Jeans, to account for its formation in this way. As is apparent, these investigations lead to conclusions very different from those derived by Chamberlin and Moulton from the planetesimal hypothesis.

This is a critical study of spiral nebulæ for which the 100-inch telescope is of all instruments in existence the best suited. The spectra of the spirals must be studied, as well as the motions of the matter composing the arms. Their parallaxes, too, must be ascertained. A photographic campaign including spiral nebulæ of various types will settle the question of internal motions. The large scale of the spiral nebulæ at the principal focus of the Hooker telescope, and the experience gained in the measurement of nebular nuclei for parallax determination, will help greatly in this research. A multiple-slit spectrograph, already applied at Mount Wilson, will be employed, not only on spiral nebulæ whose plane is directed toward us, but also on those whose plane lies at an angle sufficient to permit both components of motion to be measured by the two methods.

In dealing with problems of structure and motion in the Galactic system, the 100-inch telescope offers especial advantages, because of its vast light-gathering power. Studies of radial velocities of the stars have hitherto been necessarily confined to the brighter stars, for the most part even to those visible to the naked eye. While some of these are very distant, most of the stars whose radial velocities are known belong to a very limited group, perhaps constituting a distinct cluster of which the sun is a member, but in any event of insignificant proportions when contrasted with the Galaxy. Current spectrographic work with the 60-inch telescope includes stars of the eighth magnitude, and some even fainter. But while the 60-inch has enabled Adams to measure the distances of many remote stars by his new spectroscopic method, and to double the known extent (so far as spectroscopic evidence is concerned) of the star streams of Kapteyn, a much greater advance into space is necessary to find out the community of motion among the stars comprising the Galactic system. The Hooker telescope will enable us to determine accurate radial velocities to stars of the eleventh magnitude, which doubtless truly represent the Galaxy.

In order to secure a maximum return within a reasonable period of time, the stars in the selected areas of Kapteyn will be given the preference, because of the vast amount of work already done, relating to their positions, proper motions, and visual and photographic magnitudes. Such consideration as spectral type, the known directions of star-streaming, and the position of the chosen regions with reference to the plane of the Galaxy are given adequate weight, and it is of fundamental importance that the method of spectroscopic parallaxes will permit dwarf stars to be distinguished from stars that are in the giant class, but rendered faint by their much greater distance. In addition to these problems, the stellar spectrograms will provide rich material for study of the relationship between stellar mass and speed, and the nature of giant stars and dwarf stars.

Shapley's recent studies of globular clusters have indicated the significance of these objects in both evolutional and structural problems, and the possibility of determining their parallaxes by a number of independent methods is of prime importance, both in its bearing on the structure of the universe and because it permits a host of apparent magnitudes to be at once transformed into absolute magnitudes. Here the advantage of the Hooker telescope is two-fold: at its 134-foot focus the increased scale of the crowded clusters makes it possible to select separate stars for spectrum photography (which could not be done with the 60-inch where the images were commingled); and the great gain in light is such that the spectra of stars to the 14th magnitude have been photographed in less than an hour.

Faint globular clusters, then, will comprise a large part of the early program with the 100-inch telescope: the faintest possible stars in them must be detected and their magnitudes and colors measured; spectral types must be determined, and the radial velocities of individual stars and of clusters as a whole; spectroscopic evidence of possible axial rotation of globular clusters must be searched for; and the method of spectroscopic parallaxes, as well as other methods, must be applied to ascertaining the distances of these clusters.

The possibility of dealing with many problems relating to the distribution and evolution of the faintest stars depends upon the establishment of photographic and photovisual magnitude scales. Below the twelfth magnitude, the only existing scale of standard visual or photovisual magnitudes is the Mount Wilson sequence, already extended by Seares to magnitude 17.5 with the 60-inch telescope.

Extension of this scale to even fainter magnitudes, and its application to the faintest stars within its range is an important task for this great telescope, as it will doubtless bring within range hundreds of millions of stars that are beyond the reach of the 60-inch. The giants among them will form for us the outer boundary of the Galactic system, while the dwarfs will be of almost equal interest from the evolutional standpoint. The photometric program of the 100-inch, then, will deal with such questions as the condensation of the fainter stars toward the Galactic plane, the color of the most distant stars, and the final settlement of the long inquiry regarding the possible absorption of light in space.

Another research of exceptional promise will be undertaken, which is of great importance in a general study of stellar evolution; and that is the determination of the spectral-energy curves of stars of various classes, for the purpose of measuring their surface temperatures. A very few of the nebulæ are found to be variable, and their peculiarities need investigation, also special problems of variable stars and temporary stars, and the spectra of the components of close double stars which are beyond the power of all other instruments to photograph.

Such a program of research conveys an excellent idea of many of the great problems that are under investigation by astronomers to-day, and gives some notion of the instrumental means requisite in executing comprehensive plans of this character. It will not escape notice that the climax of instrumental development attained at Mount Wilson has only been made possible by an unbroken chain of progress, link by link, each antecedent link being necessary to the successful forging of its following one. In very large part, and certainly indispensable to these instrumental advances, has the art of working in glass and metals been the mainstay of research. As we review the history of astronomical progress, from Galileo's time to our own, the consummate genius of the artisan and his deft handiwork compel our admiration almost equally with the keen intelligence of the astronomer who uses these powerful engines of his own devising to wrest the secrets of nature from the heavens.

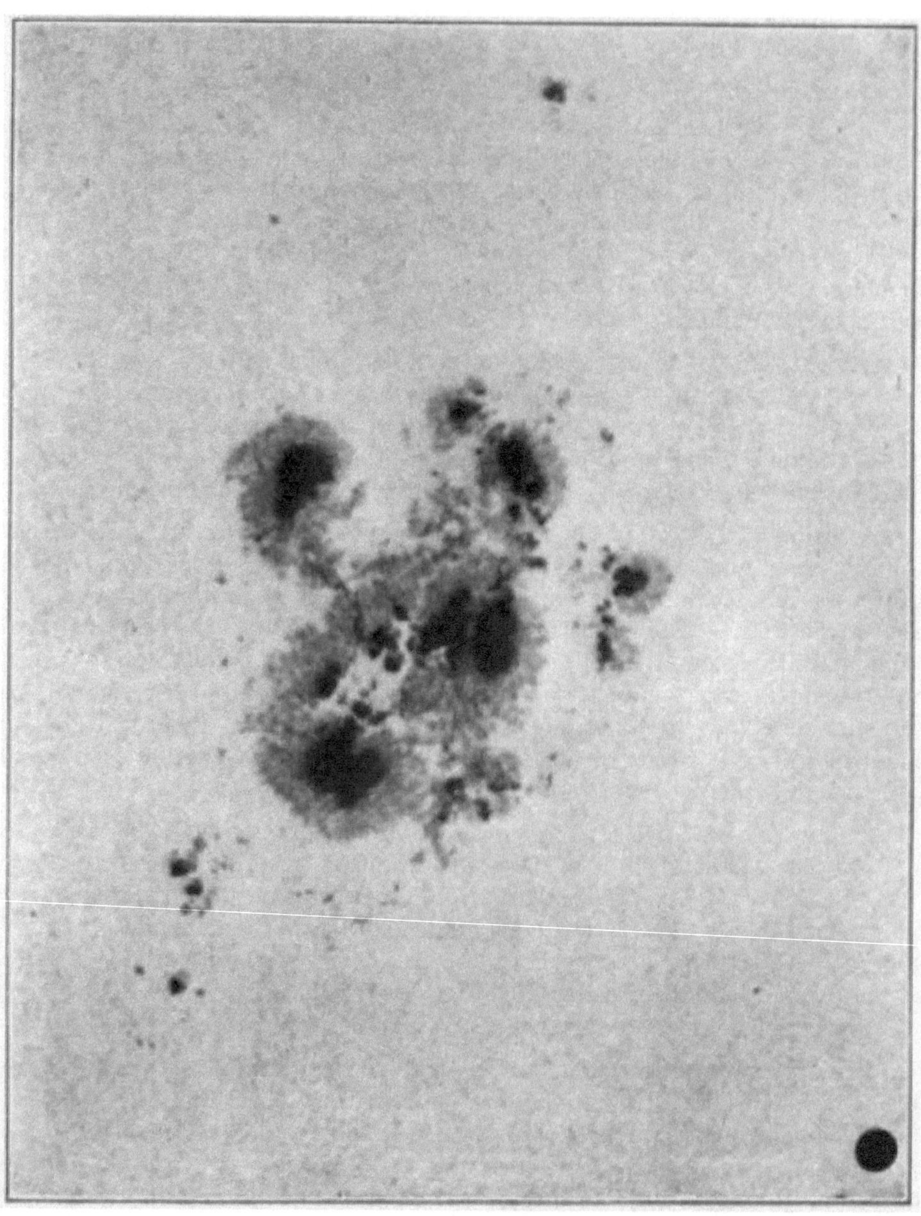

Great Sun-Spot Group, August 8, 1917. The disk in the lower left corner represents the comparative size of the earth.

(Photo, Mt. Wilson Solar Observatory.)

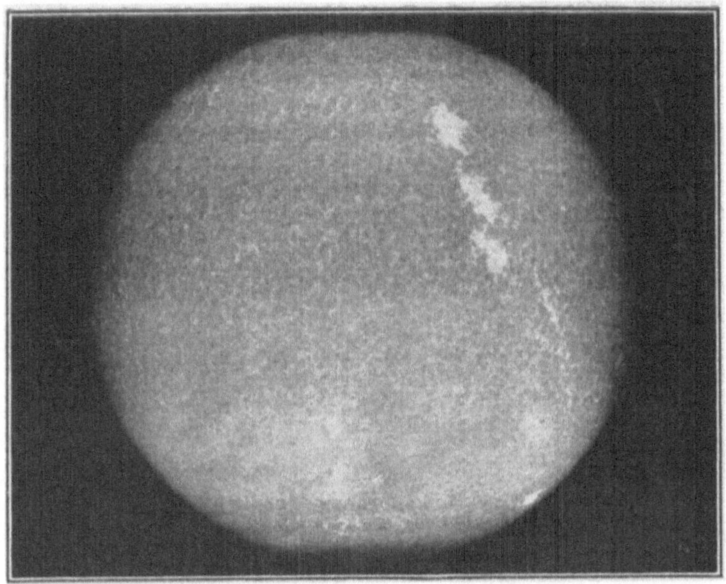

The Sun's Disk. The view shows the "rice grain" structure of the photosphere and brilliant calcium flocculi.

(Photo, Yerkes Observatory.)

The Lunar Surface Visible During a Total Eclipse of the Moon, February 8, 1906.

(Photo, Yerkes Observatory.)

CHAPTER XXIV
OUR SOLAR SYSTEM

Now let us go upward in imagination, far, far beyond the tops of the highest mountains, beyond the moon and sun, and outward in space until we reach a point in the northern heavens millions and millions of miles away, directly above and equally distant from all points in the ecliptic, or path in which our earth travels yearly round the sun. Then we should have that sort of comprehensive view of the solar system which is necessary if we are to visualize as a whole the working of the vast machine, and the motions, sizes, and distances of all the bodies that comprise it. Of such stupendous mechanism our earth is part.

Or in lieu of this, let us attempt to get in mind a picture of the solar system by means of Sir William Herschel's apt illustration: "Choose any well-leveled field. On it place a globe two feet in diameter. This will represent the sun; Mercury will be represented by a grain of mustard seed on the circumference of a circle 164 feet in diameter for its orbit; Venus, a pea on a circle of 284 feet in diameter; the Earth also a pea, on a circle of 430 feet; Mars a rather larger pin's head on a circle of 654 feet; the asteroids, grains of sand in orbits of 1,000 to 1,200 feet; Jupiter, a moderate sized orange in a circle of nearly half a mile across; Saturn, a small orange on a circle of four-fifths of a mile; Uranus, a full-sized cherry or small plum upon the circumference of a circle more than a mile and a half; and finally Neptune, a good-sized plum on a circle about two miles and a half in diameter.... To imitate the motions of the planets in the above mentioned orbits, Mercury must describe its own diameter in 41 seconds; Venus in 4 minutes, 14 seconds; the Earth in 7 minutes; Mars in 4 minutes 48 seconds; Jupiter in 2 minutes 56 seconds; Saturn in 3 minutes 13 seconds; Uranus in 2 minutes 16 seconds; and Neptune in 3 minutes 30 seconds."

Now, let us look earthward from our imaginary station near the north pole of the ecliptic. All these planetary bodies would be seen to be traveling eastward round the sun, that is, in a counter-clockwise direction, or contrary to the motions of the hands of a timepiece. Their orbits or paths of motion are very nearly circular, and the sun is practically at the center of all of them except Mercury and Mars; of Venus and Neptune, almost at the absolute center. The planes of all their orbits are very nearly the same as that of the ecliptic, or plane in which the earth moves. These and many other resemblances and characteristics suggest a uniformity of origin which comports with the idea of a family, and so the whole is spoken of as the solar system, or the sun and his family of planets.

In addition to the nine bodies already specified, the solar system comprises a great variety of other and lesser bodies; no less than twenty-six moons or satellites tributary to the planets and traveling round them in various periods as the moon does round our earth. Then between the orbits of Mars and Jupiter are many thousands of asteroids, so called, or minor planets (about 1,000 of them have actually been discovered, and their paths accurately calculated). And at all sorts of angles with the planetary orbits are the paths of hundreds of comets, delicate filmy bodies of a wholly different constitution from the planets, and which now and then blaze forth in the sky, their tails appearing much like the beam of a searchlight, and compelling for the time the attention of everybody. Connected with the comets and doubtless originally parts of them are uncounted millions of millions of meteors, which for the time become a part of the solar system, their minute masses being attracted to the planets, upon which they fall, those hitting the earth being visible to us as familiar shooting stars.

We next follow the story of astronomy through the solar system, beginning with the sun itself and proceeding outward through his family of planets, now much more numerous and vastly more extended than it was to the ancient world, or indeed till within a century and a half of our own day.

CHAPTER XXV
THE SUN AND OBSERVING IT

As lord of day, king of the heavens, mankind in the ancient world adored the sun. By their researches into the epoch of the Assyrians, Hittites, Phœnicians and other early peoples now passed from earth, archæologists have unearthed many monuments that evidence the veneration in which the early peoples who inhabited Egypt and Asia Minor many thousand years ago held the sun. A striking example is found in the architecture of early Egyptian temples, on the lintels of which are carved representations of the winged globe or the winged solar disk, and there is a bare possibility that the wings of the globe were suggested by a type of the solar corona as glimpsed by the ancients.

Little knew they about the distance and size of the sun; but the effects of his light and heat upon all vegetal and animal life were obvious to them. Doubtless this formed the basis for their worship of the sun. Occasional huge spots must have been visible to the naked eye, and the sun's corona was seen at rare intervals. Plutarch and Philostratus describe it very much as we see it to-day.

How completely dependent mankind is upon the sun and its powerful radiations, only the science of the present day can tell us. By means of the sun's heat the forests of early geologic ages were enabled to wrest carbon from the atmosphere and store it in forms later converted by nature's chemistry into peat and coal. Through processes but imperfectly understood, the varying forms of vegetable life are empowered to conserve, from air and soil, nitrogen and other substances suitable for and essential to the life maintenance of animal creatures. Breezes that bring rain and purify the air; the energy of water held under storage in stream and dam and fall; trade winds facilitating commerce between the continents; oceanic currents modifying coastal climates; the violence of tornado, typhoon and water-spout, together with other manifestations of natural forces— all can be traced back to their origin in the tremendous heating power of the solar rays. In everything material the sun is our constant and bountiful benefactor. If his light and heat were withdrawn, practically every form of human activity on this planet would come to an early end.

How far away is the sun? What is the size of the sun? These are questions that astronomers of the present day can answer with accuracy.

So closely do they know the sun's distance that it is employed as their yardstick of the sky, or unit of celestial measurement. Many methods have been utilized in ascertaining the distance of the sun, and the remarkable agreement among them all is very extraordinary. Some of them depend upon pure geometry, and the basic measure which we make from the earth is not the distance of the sun directly; but we find out how far away Venus is during a transit of Venus, for example, or how far away Mars is or some of the asteroids are at their closer oppositions. Then it is possible to calculate how far away the sun is, because one measurement of distance in the solar system affords us the scale on which the whole structure is built. But perhaps the simplest method of getting the sun's distance is by the velocity of light, 186,300 miles a second. From eclipses of Jupiter's moons we know that light takes 8 minutes 20 seconds to pass from sun to earth. So that the sun's distance is the simple product of the two, or 93 millions of miles.

Once this fundamental unit is established, we have a firm basis on which to build up our knowledge of the distances, the sizes and motions of the heavenly bodies, especially those that comprise the solar system. We can at once ascertain the size of the sun, which we do by measuring the angle which it fills, that is, the sun's apparent diameter. Finding this to be something over a half a degree in arc, the processes of elementary trigonometry tell us that the sun's globe is 865,000 miles in diameter. For nearly a century this has been accurately measured with the greatest care, and diameters taken in every direction are found to be equal and invariably the same. So we conclude that the sun is a perfect sphere, and so far as our instruments can inform us, its actual diameter is not subject to appreciable change.

The vastness of the sun's volume commands our attention. As his diameter is 110 times that of the earth, his mere size or volume is 110×110×110 or 1,300 thousand times that of the earth, because the volumes of spheres are in proportion as the cubes of their diameters. If the materials that compose the sun were as heavy as those that make up the earth, it would take 1,300 thousand earths to weigh as much as the sun does. But by a method which we need not detail here, the sun's actual weight or mass is found to be only 300 thousand (more nearly 330,000), times greater than the earth's. So we must infer that, bulk for bulk, the component materials of the sun are about one-fourth lighter than those of the earth, that is, about one and one-half times as dense as water.

To look at this in another way: it is known that a body falling freely toward the earth from outer space would acquire a speed of seven miles a second, whereas if it were to fall toward the sun instead, the velocity would be 383 miles a second on reaching his surface. If all the other bodies of the solar system, that is, the earth and moon, all the planets and their satellites, the comets and all were to be fused together in a single globe, it would weigh only one-seven hundred and fiftieth as much as the sun does.

At the surface, however, the disproportion of gravity is not so great, because of the sun's vast size: it is only about twenty-eight times greater on the sun than on the earth; and instead of a body falling 16 feet the first second as here, it would fall 444 feet there. Pendulums of clocks on the sun would swing five times for every tick here, and an athlete's running high jump would be scaled down to three inches.

Let us next inquire into the amount of the sun's light and heat, and the enormously high temperature of a body whose heat is so intense even at the vast distance at which we are from it. The intensity of its brightness is such that we have no artificial source of light that we can readily compare it with. In the sky the next object in brightness is the full moon, but that gives less than the half-millionth part as much light as the sun. The standard candle used in physics gives so little light in comparison that we have to use an enormous number to express the quantity of light that the sun gives.

A sperm candle burning 120 grains hourly is the standard, and if we compare this with the sun when overhead, and allow for the light absorbed by the atmosphere, we get the number 1575 with twenty-four ciphers following it, to express the candlepower of the sun's light. If we interpose the intense calcium light or an electric arc light between the eye and the sun, these artificial sources will look like black spots on the disk. Indeed, the sun is nearly four times brighter than the "crater," or brightest part of the electric arc. The late Professor Langley at a steel works in Pennsylvania once compared direct sunlight with the dazzling stream of molten metal from a Bessemer converter; but bright as it was, sunlight was found to be five thousand times brighter.

Equally enormous is the heat of the sun. Our intensest sources of artificial heat do not exceed 4,000 degrees Fahrenheit, but the temperature at the sun's surface is probably not less than 16,000 degrees F. One square meter of his surface radiates enough heat to generate 100,000 horsepower continuously. At our vast distance of 93 millions of miles, the sun's heat received by the earth is still powerful enough to melt annually a layer of ice on the earth more than a hundred feet in thickness. If the solar heat that strikes the deck of a tropical steamship could be fully utilized in propelling it, the speed would reach at least ten knots.

Many attempts have been made in tropical and sub-tropical climates to utilize the sun's heat directly for power, and Ericsson in Sweden, Mouchot in France, and Shuman in Egypt have built successful and efficient solar engines. Necessary intermission of their power at night, as well as on cloudy days, will preclude their industrial introduction until present fuels have advanced very greatly in cost. All regions of the sun's disk radiate heat uniformly, and the sun's own atmosphere absorbs so much that we should receive 1.7 times more heat if it were removed. So far as is known, solar light and heat are radiated equally in all directions, so that only a very minute fraction of the total amount ever reaches the earth, that is, 1 2200 millionth part of the whole. Indeed all the planets and other bodies of the solar system together receive only one one hundred millionth part; the vast remainder is, so far as we know, effectively wasted. It is transformed, but what becomes of it, and whether it ever reappears in any other form, we cannot say.

How is this inconceivably vast output of energy maintained practically invariable throughout the centuries? Many theories have been advanced, but only one has received nearly universal assent, that of secular contraction of the sun's huge mass upon itself. Shrinkage means evolution of heat; and it is found by calculation that if the sun were to contract its diameter by shrinking only two-hundred and fifty feet per year, the entire output of solar heat might thus be accounted for. So distant is the sun and so slow this rate of contraction that centuries must elapse before we could verify the theory by actual measurements. Meanwhile, the progress of physical research on the structure and elemental properties of matter has brought to light the existence of highly active internal forces which are doubtless intimately concerned in the enormous output of radiant energy, though the mechanism of its maintenance is as yet known only in part.

Abbot, from many years' observations of the solar constant, at Washington, on Mount Wilson, and in Algeria, finds certain evidence of fluctuation in the solar heat received by the earth. It cannot be a local phenomenon due to disturbances in our atmosphere, but must originate in causes entirely extraneous to the earth. Interposition of meteoric dust might conceivably account for it, but there is sufficient evidence to show that the changes must be attributed to the sun itself. The sun, then, is a variable star; and it has not only a period connected with the periodicity of the sun spots, but also an irregular, nonperiodic variation during a cycle of a week or ten days, though sometimes longer, and occasioning irregular fluctuations of two to ten per cent of the total radiation. Radiation is found to increase with the spottedness.

Attempts have been made on the basis of the contraction theory to find out the past history of the sun and to predict its future. Probably 20 to 50 millions of years in the past represents the life of the sun much as it is at present; and if solar radiation in the future is maintained substantially as now, the sun will have shrunk to one-half its present diameter in the next five million years.

So far then as heat and light from the sun are concerned, the sun may continue to support life on the earth not to exceed ten million years in the future. But the sun's own existence, independently of the orbs of the system dependent upon it, might continue for indefinite millions of aeons before it would ever become a cold dead globe; indeed, in the present state of science, we cannot be sure that it is destined to reach that condition within calculable time.

A few words on observing the sun, an object much neglected by amateurs. On account of the intense light, a very slight degree of optical power is sufficient. Indeed a piece of window glass, smoked in a candle flame with uniform graduation from end to end, will be found worth while in a beginner's daily observation of the sun. The glass should be smoked densely enough at one end so that the sunlight as seen through it will not dazzle the eye on the clearest days. At the other end of the glass, the degree of smoke film should not be quite so dense, so that the sun can be examined on hazy, foggy or partly cloudy days. An occasional naked-eye spot will reward the patient observer.

If a small spyglass, opera glass or field glass is at hand, excellent views of the sun may be had by mounting the glass so that it can be held steadily pointed on the sun, and then viewing the disk by projection on a white card or sheet of paper. Care must be taken to get a good focus on the projected image, and then the faculæ, or whitish spots, or mottling nearer the sun's edge will usually be well seen. By moving the card farther away from the eyepiece, a larger disk may be obtained, in effect a higher degree of magnification. But care must be used not to increase it too much. Keep direct sunlight outside the tube from falling on the card where the image is being examined. This is conveniently done by cutting a large hole, the size of the brass cell of the object glass, through a sheet of corrugated strawboard, and slipping this on over the cell. In this way the spots on the sun can be examined with ease and safety to the eye.

For large instruments a special type of eyepiece is provided known as a helioscope, which disposes of the intense heat rays that are harmful to the eye. Frequent examination of the eyepiece should be made and the eyepiece cooled if necessary. That part of the sun's surface under observation is known as the photosphere, that is, the part which radiates light. If the atmosphere admits the use of high magnifying powers, the structure of the photosphere will be found more and more interesting the higher the power employed. It is an irregularly mottled surface showing a species of rice-grain structure under fairly high magnification. These grains are grouped irregularly and are about 500 miles across. Under fine conditions of vision they may be subdivided into granules. The faculæ, or white spots, are sometimes elevations above the general solar level; they have occasionally been seen projecting outside the limb, or edge of the disk.

CHAPTER XXVI
SUN SPOTS AND PROMINENCES

Dark spots of a deep bluish black will often be seen on the photosphere of the sun. Sometimes single, though generally in groups, the larger ones will have a dark center, called the umbra, surrounded by the very irregular penumbra which is darker near its outer edge and much brighter apparently on its inner edge where it joins on the umbra. The penumbra often shows a species of thatch-work structure, and systematic sketches of sun spots by observers skilled in drawing are greatly to be desired, because photography has not yet reached the stage where it is possible to compete with visual observation in the matter of fine detail. The spots themselves nearly always appear like depressions in the photosphere, and on repeated occasions they have been seen as actual notches when on the edge of the sun.

Many spots, however, are not depressions: some appear to be actual elevations, with the umbra perhaps a central depression, like the crater in the general elevation of a volcano. Spots are sometimes of enormous size. The largest on record was seen in 1858; it was nearly 150,000 miles in breadth, and covered a considerable proportion of the whole visible hemisphere of the sun. A spot must be nearly 30,000 miles across in order to be seen with the naked eye.

In their beginning, development, and end, each spot or group of spots appears to be a law unto itself. Sometimes in a few hours they will form, though generally it is a question of days and even weeks. Very soon after their formation is complete, tonguelike encroachments of the penumbra appear to force their way across the umbra, and this splitting up of the central spot usually goes on quite rapidly. Sun spots in violent disturbance are rarely observed. As the sun turns round on his axis, the spots will often be carried across the disk from the center to the edge, when they become very much foreshortened. The sun's period of rotation is 28 days, so that if a spot lasts more than two weeks without breaking up, it may reappear on the eastern limb of the sun after having disappeared at the western edge. Two or three months is an average duration for a spot; the longest on record lasted through 18 months in 1840-41.

The position of the sun's axis is well known, its equator being tilted about 7 degrees to the ecliptic, and the spots are distributed in zones north and south of the equator, extending as far as 30 degrees of solar latitude. In very high latitudes spots are never seen; they are most abundant in about latitude 15 degrees both north and south, and rather more numerous in the northern than in the southern hemisphere of the sun. Recent research at Mount Wilson makes the sun a great magnet; and its magnetic axis is inclined at an angle of 6 degrees to the axis of rotation, around which it revolves in 32 days.

There is a most interesting periodicity of the spots on the sun, for months will sometimes elapse with spots in abundance and visible every day, while at other periods, days and even weeks will elapse without a single spot being seen. There is a well recognized period of eleven and one-tenth years, the reason underlying which is not, however, known. After passing through the minimum of spotted-ness, they begin to break out again first in latitudes of 25 degrees-30 degrees, rather suddenly, and on both sides of the equator, and they move toward the equator as their number and individual size decrease.

The last observed epoch of maximum spot activity on the sun was passed in 1917.

Many attempts have been made to ascertain the cause of the periodicity of sun spots, but the real cause is not yet known. If the spots are eruptional in character, the forces held in check during seasons of few spots may well break out in period. The brighter streaks and mottlings known as faculæ are probably elevations above the general photosphere, and seem to be crusts of luminous matter, often incandescent calcium, protruding through from the lower levels. Generally the faculæ are numerous around the dark spots, and absorption of the sun's light by his own atmosphere affords a darker background for them, with better visibility nearer the rim of the solar disk. The spectroheliograph reveals vast zones of faculæ otherwise invisible, related to the sun-spot zones proper on both sides of the equator.

In some intimate way the magnetism of sun and earth are so related that out-breaks of solar spots are accompanied with disturbances of electrical and other instruments on the earth; also the aurora borealis is seen with greater frequency during periods when many spots are visible.

Within very recent years the discovery of a magnetic field in sun spots has been made by Hale with powerful instruments of his own design. Sun spots had never been investigated before with adequate instrumental means. He recognized the necessity of having a spectroscope that would record the widened lines of sun-spot spectra, and the strengthened and weakened lines on a large scale. Certain changes in relative intensity were traced to a reduced temperature of the spot vapors by comparison with photographs of the spectrum of iron and other metallic vapors in an electric arc at different temperatures. Here the work of the laboratory was essential. Sun spots were thus found to be regions of reduced temperature in the solar atmosphere. Chemical unions were thus possible, and thousands of faint lines in spot-spectra were measured and identified as band lines due to chemical compounds. Thus the chemical changes at work in sun-spot vapors were recognized.

Then followed the highly significant investigations of solar vortices and magnetic fields. Improvements in photographic methods had revealed immense vortices surrounding sun spots in the higher part of the hydrogen atmosphere; and this led to the hypothesis that a sun spot is a solar storm, resembling a terrestrial tornado, and in which the hot vapors whirling at high velocity are cooled by expansion. This would account for the observed intensity changes of the spectrum lines and the presence of chemical compounds. The vortex hypothesis suggested an explanation of the widening of many spot lines, and the doubling or trebling of some of them. As it is known that electrons are emitted by hot bodies, they must be present in vast numbers in the sun; and positive or negative electrons, if caught and whirled in a vortex, would produce a magnetic field.

Zeeman in 1896 had discovered that the lines in the spectrum of a luminous vapor in a magnetic field are widened, or even split into several components if the field is strong enough. Characteristic effects of polarization appear also. The new apparatus of the observatory in conjunction with experiments in the laboratory immediately provided evidence that proved the existence of magnetic fields in sun spots, and strengthened the view that the spots are caused by electric vortices.

Extended investigations have led Hale to the conclusion that the sun itself is a magnet, with its poles situated at or near the poles of rotation. In this respect the sun resembles the earth, which has long been known to be a magnet. The sun's axial rotation permits investigation of the magnetic phenomena of all parts of its surface, so that ultimately the exact position of the sun's magnetic poles and the intensity of the field at different levels in the solar atmosphere will be ascertained. Schuster is of the opinion that not only the sun and earth, but every star, and perhaps every rotating body, becomes a magnet by virtue of its rotation. Hale is confident that the 100-inch reflector will permit the test for magnetism to be applied to a few of the stars.

The sun can be observed at Mount Wilson on at least nine-tenths of all the days in the year, and a daily record of the polarities of all spots with the 150-foot tower telescope is a part of the routine. A method has been devised for classifying sun spots on the basis of their magnetic properties, and more than a thousand spots have already been so classified. About 60 per cent of all sun spots are found to be binary groups, the single or multiple members of which are of opposite magnetic polarity. Unipolar spots are very seldom observed without some indication of the characteristics of bipolar groups. These are usually exhibited in the form of flocculi following the spot. The bipolar spot seems to be the dominant type, and the unipolar type a variant of it.

Although devised for quite another purpose, that of photographing the hydrogen prominences on the limb of the sun, the spectroheliograph has contributed very effectively to many departments of solar research. The prominences are dull reddish cloudlets that were first seen during total eclipses of the sun. Probably Vassenius, a Swedish astronomer, during the total eclipse of 1733, made the earliest record of them, as pinkish clouds quite detached from the edge of the moon; and in that day, when it had not yet been proved that the moon was without atmosphere, he naturally thought they belonged to the moon, not the sun. Undoubtedly Ulloa, a Spanish admiral, also saw the prominences in observing the total eclipse of 1778; but they seem to have attracted little attention till 1842, when a very important total eclipse was central throughout Europe, and observed with great care by many of the eminent astronomers of all countries.

So different did the prominences appear to different eyes, and so many were the theories as to what they were, that no general consensus of opinion was reached, and some thought them no part of either sun or moon, but a mere mirage or optical illusion. But at the return of this eclipse in 1860, photography was employed so as to demonstrate beyond a shadow of doubt the real existence and true solar character of the prominences. By the slow progress of the moon across the sun and the prominences on the edge, a unique series of photographs by De la Rue showed the moon's edge gradually cutting off the prominences piecemeal on one side of the sun, and equally gradually uncovering them on the opposite side.

The prominences, then, were known to be real phenomena of the sun, some of them disconnectedly floating in his atmosphere, as if clouds. Their forms did not vary rapidly, they were very abundant, and their light was so rich in rays of great photographic intensity that many were caught on the plate which the eye failed to see; they appeared at every part of the sun's limb and their height above it indicated that they must be many thousand miles in actual dimension. What they were, however, remained an entire mystery, and no one even thought it possible to find out what their chemical constitution might be or to measure the speed with which they moved.

A few years later came the great Indian eclipse (August 28, 1868), at that date the longest total eclipse ever observed. Janssen of France and many others went out to India to witness it. Fortunately the prominences were very brilliant and this led Janssen to believe it would be possible for him to see them the day after the eclipse was over. By modifying the adjustment of his apparatus suitably and changing its relation to the sun's edge, he found that hydrogen is the main constituent in the light of the prominences. In addition to this he was able to trace out the shapes of the prominences, and even measure their dimensions. His station in India was at Guntoor, many weeks by post from home; so that his account of this important discovery reached the Paris Academy of Sciences for communication with another from the late Sir Norman Lockyer of England, announcing a like discovery, wholly independently.

The principle is simply this, and admirably stated by Young: "Under ordinary circumstances the prominences are invisible, for the same reason as the stars in the daytime: they are hidden by the intense light reflected from the particles of our own atmosphere near the sun's place in the sky; and if we could only sufficiently weaken this aerial illumination, without at the same time weakening *their* light, the end would be gained. And the spectroscope accomplishes this very thing. Since the air-light is reflected sunshine, it of course presents the same spectrum as sunlight, a continuous band of color crossed by dark lines. Now, this sort of spectrum is greatly weakened by every increase of dispersive power, because the light is spread out into a longer ribbon and made to cover a more extended area. On the other hand, a spectrum of bright lines undergoes no such weakening by an increase in the dispersive power of the spectroscope. The bright lines are only more widely separated—not in the least diffused or shorn of their brightness."

Simultaneous announcement of this great discovery, by astronomers of different nations, working in widely separate regions of the earth, led to the striking of a gold medal by the French Government in honor of both astronomers and bearing their united effigies. Ever since the famous Indian eclipse of 1868, it has not been necessary to wait for a total eclipse in order to observe the solar prominences, but every observer provided with suitable apparatus has been able to observe them in full sunlight whenever desired, and the charting of them is part of the daily routine at several observatories in different parts of the world. So vast has been the accumulation of data about them that we know their numbers to fluctuate with the spots on the sun; and their distribution over the sun's surface resembles in a way that of the spots.

While the spots and protuberances are most numerous around solar latitude 20 degrees both north and south, the prominences do not disappear above latitude 35 to 40 degrees, as the spots do, but from latitude 60 degrees they increase in number to about 75 degrees, and are occasionally observed even at the sun's poles. Faculæ and prominences are more closely related than the sun spots and prominences. There are wide variations in both magnitude and type of the prominences. Heights above the sun's limb of a few thousand miles are very common, and they rarely reach elevations as great as 100,000 miles, though a very occasional one reaches even greater heights.

Classification of the prominences divides them into two broad types, the quiescent and the eruptive. The former are for the most part hydrogen, and the latter metallic. The quiescent prominences resemble closely the stratus and cirrus type of terrestrial clouds, and are frequently of enormous extent along the sun's edge. They are relatively long-lived, persisting sometimes for days without much change. The eruptive prominences are more brilliant, changing their form and brightness rapidly. Often they appear as brilliant spikes or jets, reaching altitudes that average about 25,000 miles. Rarely seen near the sun's poles, they are much more numerous nearer the sun spots. Speed of motion of their filaments sometimes exceeds one hundred miles a second, and the changing variety of shapes of the eruptive prominences is most interesting. Oftentimes they change so rapidly that only photography can do them justice.

Prominence photography began with Young a half century ago, who obtained the first successful impression on a microscope slide with a sensitized film of collodion; as was necessary in the earlier wet-plate process of photography, which required exposures so long that little progress was effected for about twenty years. Then it was taken up by Deslandres of Paris and Hale of Chicago independently, both of whom succeeded in devising a complex type of apparatus known as the spectroheliograph, by which all the prominences surrounding the entire limb of the sun can be photographed at any time by light of a single wave-length, together with the disk of the sun on the same negative.

The prominences appear to be intimately connected with a gaseous envelope surrounding the solar photosphere, in which sodium and magnesium are present as well as hydrogen. The depth of the chromosphere is usually between 5,000 and 10,000 miles, and its existence was first made out during the total solar eclipses of 1605 and 1706, when it appeared as an irregular rose-tinted fringe, though not at the time recognized as belonging to the sun.

The constitution of the sun and its envelopes are still under discussion, and no complete theory of the sun has yet been advanced which commands the widest acceptance. Of the interior of the sun we can only surmise that it is composed of gases which, because of intense heat and compression, are in a state unfamiliar on earth and impossible to reproduce in our laboratories. Their consistency may be that of melted pitch or tar.

Surrounding the main body of the sun are a series of layers, shells, or atmospheres. Outside of all and very irregular in structure, indeed probably not a solar atmosphere at all, is the solar corona, parts of which behave much as if it were an atmosphere, but it appears to be bound up in some way with the sun's radiation. It has streamers that vary with the sun-spot period, but its constitution and function are very imperfectly known, because it has never been seen or photographed except at rare intervals on occasion of total eclipses of the sun.

Beneath the corona we meet the projecting prominences, to which parts of the corona are certainly related, and beneath them the first true layer or atmosphere of the sun known as the chromosphere, its average depth being about one-hundredth part of the sun's diameter. Beneath the chromosphere is the layer of the sun from which emanates the light by which we see it, called the photosphere. It appears to be composed of filaments due to the condensation of metallic vapors, and it is the outer extremities of these filaments which are seen as the granular structures everywhere covering the disk of the sun. Their light shines through the chromosphere and the spots are ruptures in this envelope.

Between photosphere and chromosphere is a very thin envelope, probably not over 700 miles in thickness, called the reversing layer. It is this relatively thin shell that is responsible for the absorption which produces the dark lines in the spectrum of the sun. Under normal conditions the filaments of the photosphere are radial, that is vertical on the sun; but whenever eruptions take place, as during the occurrence of spots, the adjacent filaments are violently swept out of their normal vertical lines and these displaced columns then form what we view as the spot's penumbra. From the outer surface of the sun's chromosphere rise in eruptive columns vapors of hydrogen and the various metals of which the sun is composed. These and the spots would naturally occur in periods just as we see them.

We have said that the sun is composed of a mass of highly heated or incandescent vapors or gases, whose compression on account of gravity must render their physical condition quite different from any gaseous forms known on the earth or which we can reproduce here. As the result of more than half a century of studious observation of the sun and mapping of its spectrum in every part, and diligent comparison with the spectra of all known chemical elements on the earth, we find that the sun contains no elements not already found here, but that a great preponderance of elements known to earth are found in the sun.

The intensity of their spectral lines is one prominent indication of the presence of elements in the sun, and the number of coincidences of spectral lines is another. Iron, nickel, calcium, manganese, sodium, cobalt, and carbon are among the elements most strongly identified. A few of the rarer terrestrial elements are of doubtful existence in the sun, and a very few, as gold, bismuth, antimony, and sulphur are not found there, and the existence of oxygen in the sun is regarded by some experts as doubtful. But if the whole earth were vaporized by heat, probably its spectrum would resemble that of the sun very closely.

What are the effects of the sun, and sun spots in particular, on our weather? Is the influence of their periodicity potent or negligible? If we investigate conditions pertaining to terrestrial magnetism, as fluctuations of the magnetic needle, and the frequency of auroræ, there is no occasion for doubt of the sun's direct influence, although we are not able to say just how that influence becomes potent. If, however, we look into questions of temperature, barometric pressure, rainfall, cyclones, crops, and consequent financial conditions, we find fully as much evidence against solar influence as for it. The slight variations of the sun's light and heat due to the presence or absence of sun spots can scarcely be sensible, and much longer periods of closer observation are necessary before such questions can be finally decided. The slighter such influences are, if they actually exist, and the more veiled they are by other influences more or less powerful, the more difficult it is to discover their effects with certainty.

The importance of solar radiation in the prediction of terrestrial weather has long been recognized, but until very recently no practical application has been made. The Smithsonian Astrophysical Observatory at Washington, under the direction of Dr. Abbot, has for many years carried on at a number of stations a series of determinations of the constant of solar radiation by the spectro-bolometric method originated by Langley. A new station in Calama, Chile, has recently been inaugurated, at which the solar constant is worked out each day, and telegraphed to the Argentine weather service, where it is employed in forecasting for the day.

Abbot's new method of solar constant determination is based on the fact that atmospheric transparency varies oppositely to the variations of brightness of the sky. Increase of haziness presents more reflecting surface to scatter the solar rays indirectly to the earth. Of course it presents also additional surface to obstruct the direct rays from the sun. By measuring the brightness of the sky near the sun, it becomes possible to infer the coefficients of atmospheric transmission at all wave lengths. The direct observations and the complete deduction of the solar constant for the day can all be completed within two or three hours.

Clayton of Buenos Aires has now employed these results in the Argentine weather predictions for two years, and the introduction of this new element in forecasting has brought about a pronounced gain in the value of the predictions. Its adoption by the weather bureaus of other nations will doubtless come in due time, and the new method take a firmly established rank in practical meteorology.

Abbot's observations many years ago first called attention to the variability of the solar constant through a range of several per cent both from year to year, and in irregular short periods of weeks or even days. Abbot considers this the more likely explanation than that atmospheric changes should take place simultaneously all over the earth. The sun is but a star, the stars that are irregularly variable in light and heat are numerous, and the sun itself appears to be one of these.

Especially important to the agricultural and vineyard interests of Argentina is the question of precipitation, and Clayton finds this very dependent on solar radiation. At epochs of practically stationary solar intensity, there is little or no precipitation; but quite generally he finds that great decrease of solar radiation is followed in from three to five days by heavy precipitation. Direct temperature effects are also traced in Buenos Aires and other South American cities, lagging from two to three days behind the observed solar fluctuations.

The station at Calama yields about 250 determinations of the solar constant each year, and the Mount Wilson station about half that number. They are the only stations of this character at present in existence, and others should be established in widely separated and cloudless regions, as Egypt, southern California and Australia. Uniformity in the methods of observing would be highly desirable, and the Smithsonian Institution has perfected the details of common control of such stations which it is expected may be established at an early day.

CHAPTER XXVII
THE INNER PLANETS

VULCAN

About the middle of the last century, Le Verrier, a great French astronomer, having added the planet Neptune beyond the outside confines of the solar system, sought evidence of a lesser planet traveling round the sun within the orbit of Mercury. For many years close watch was kept on the sun in the hope of discovering such a body in the act of passing across the disk, or in transit, as it is technically termed. Lescarbault, a French physician, announced that he had actually seen such a planet, Vulcan it was called, passing over the sun in 1859. Total eclipses of the sun would afford the best opportunity for seeing such a body, and on several such occasions astronomers thought they had found it. But the signal advantages of photography have been applied so often to this search, and always unsuccessfully, that the existence of Vulcan, or the intramercurian planet, is now regarded as mythical.

MERCURY

This planet is an elusive body that very few, even astronomers, have ever seen. It is not very bright, has a rapid motion and never retreats far from the sun, so that it was a puzzle to the ancients who saw it, sometimes in the twilight after sunset and again in the twilight of dawn. When following the sun down in the west, in March or April, Mercury is likely to be best seen; twinkling rather violently and nearly as bright as a star of the first magnitude.

Very little is to be seen on the minute disk of this planet, except that it goes through all the phases of the moon—crescent, gibbous, full, gibbous, crescent. Whether Mercury turns round on its axis or not, cannot be said to be known, because the markings that are suspected on its surface are too indefinite to permit exact observation. More than likely the planet presents always the same side or face to the sun, so that it turns round on its axis once, while traveling once around the sun in its orbit. Mercury's day and year would therefore be equal in length. Nor have we much evidence on the question of an atmosphere surrounding Mercury; probably it is very thin, if indeed there is any at all. When Mercury comes directly between us and the sun, crossing in transit, the edge of the planet as projected against the sun is very sharply defined, and this would indicate an absence of atmosphere on Mercury.

Transits of Mercury can occur in May and November only: there was one on November 7, 1914, and there will be one on May 7, 1924. The latter will be nearly eight hours in length, which is almost the limit. Mercury's distance from the sun averages 36 million miles, the diameter of the planet is 3,000 miles, and his orbital speed is 30 miles per second, the swiftest of all the planets. No moon of Mercury is known to exist, although many times diligently searched for, especially during transits of the planet.

VENUS

Brightest of all the planets, and the most beautiful of all is Venus. Its path is next outside the orbit of Mercury, but within that of the earth, so that it partakes of all the phases of the moon. Like Mercury it sometimes passes exactly between us and the sun, a rare phenomenon which is known as a transit of Venus.

Being without telescopes, the ancients knew nothing about these occurrences, but they were puzzled for centuries over the appearance of the planet in the west after sunset, when they called it Hesperus, and in early dawn in the east when they gave it the name Phosphorus.

Venus is known to be girdled with an atmosphere denser than ours, and it seems to be always filled with dense clouds. It is the reflection of sunlight from this perpetually cloudy exterior which gives Venus her singular radiance. So brilliant is she that even full daylight is not strong enough to overpower her rays; and she may often be seen glistening in the clear blue daytime sky, if one knows pretty nearly in what direction to look for her.

Venus is 67 million miles from the sun, and as our own distance is 93 million miles, this planet can come within 26 million miles of the earth. It is therefore at times our nearest known neighbor in space, excepting only the Moon and Eros, one of the erratic little planets that travel round the sun between Mars and Jupiter. Also possibly a comet might come much nearer.

Astronomers always take advantage of this nearness of Venus to us, if a transit across the sun takes place; because it affords an excellent method of finding out what the distance of the sun is from the Earth. A pair of these transits happens about once a century, there were transits in 1874 and 1882, and the next pair occur in 2004 and 2012. In actual size, Venus is almost as large a planet as our own, being 7,700 miles in diameter, as compared with 7,920 for the earth. Her velocity in her orbit is twenty-two miles per second, and she travels all the way round the sun in seven and one half months or 225 days.

Venus from her striking brilliancy always leads the novice to expect to see great things on applying the telescope. But aside from a brilliant disk, now a slender crescent, now half full like the moon at quarter, and again gibbous as the moon is between quarter and full, the telescope reveals but little. There is pretty good evidence that the markings thought to have been seen on the planet's surface are illusory, and so it is wholly uncertain in what direction the planet's axis lies; also there is great uncertainty about the length of the day on Venus, or the period of turning round on its axis. Probably it is the same in length as the planet's year.

Once when Venus passed very close to the sun, just barely escaping a transit, Lyman of Yale University caught sight of it by hiding the sun behind a tall building or church spire. The dark side of Venus was turned toward us and he could not of course see that. But the planet was clearly there, completely encircled by a narrow delicate luminous ring, which was due to sunlight shining through the atmosphere that surrounds the planet. Similar ring effects were seen by observers of the transits of Venus in 1874 and 1882; and from all their observations it is concluded that Venus has an atmosphere probably at least twice as dense and extensive as that which encircles the earth. Spurious satellites of Venus are many, but no real moon is known to attend this planet.

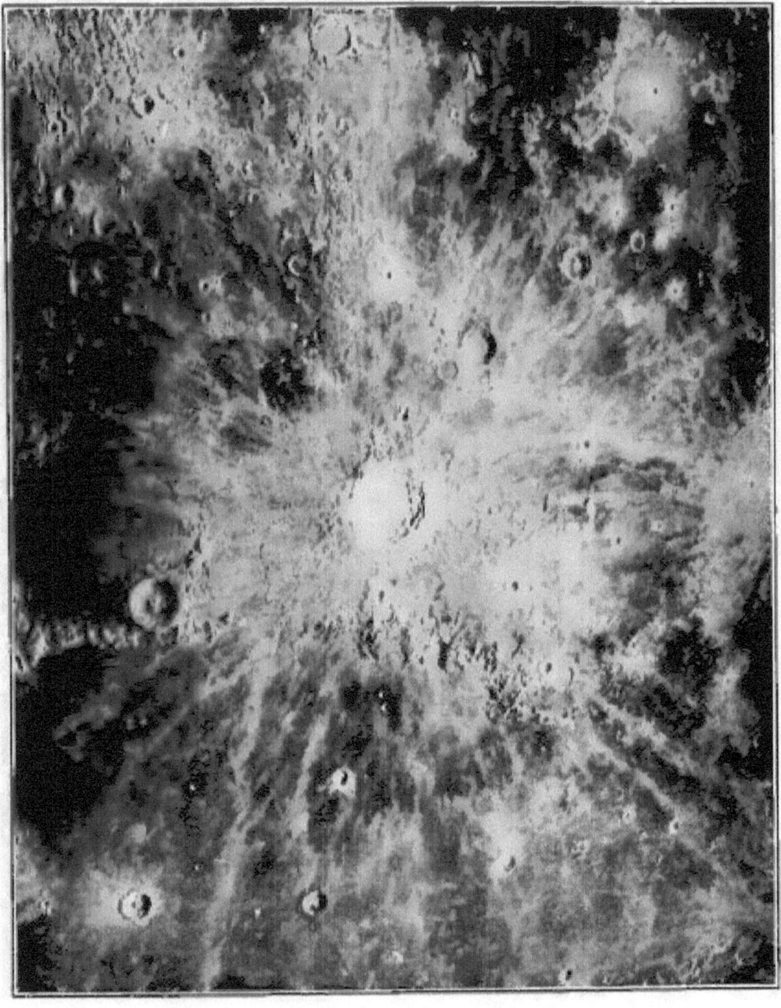

The Surface of the Moon in the Region of Copernicus. Photograph made with the Hooker 100-inch reflecting telescope.

(Photo, Mt. Wilson Solar Observatory.)

A View of the South Central Portion of the Moon at Last Quarter.

(Photo, Mt. Wilson Solar Observatory.)

CHAPTER XXVIII
THE MOON AND HER SURFACE

As the sun has always reigned as king of day, so is the moon queen of night. Observation of her phases, now waxing, now waning, with her stately motion always eastward among the stars, began with the earliest ages. Often when near the full she must have been seen herself eclipsed, and much more rarely the occurrence of total eclipses of the sun are certain to have suggested the moon's intervention between earth and sun, shutting off the sunlight completely, because these eclipses never took place except when the moon was in the same part of the sky with the sun.

If we watch the nightly march of the moon, we shall find that she travels over her own breadth in about an hour's time. By using a telescope on the stars just eastward or to the left of her, she will now and then be seen to pass between us and a star—on very rare occasions a planet—extinguishing its light with great suddenness, the most nearly instantaneous of all phenomena in nature. Draw a line connecting the cusps, or horns of the lunar crescent, and then a line eastward at right angles to this, and it will show the direction of the moon's own motion in its orbit round the earth quite accurately.

As the phase advances, note the inside edge of the advancing crescent: this will be quite rough and jagged, compared to the outside edge which is the moon's real contour and relatively very smooth. The position of the inside curve will change from night to night, and it marks the line of sunrise on the moon during the fortnight elapsing between new moon and full; while from full through last quarter and back to new moon, this advancing line marks the region of sunset on the moon. The general shape of this line is never a circle but always elliptical, and astronomers call it the terminator. All along the terminator, sunlight strikes the lunar surface at a small angle, whether near sunrise or sunset; so that owing to the mountains and other high masses of the moon's surface, the terminator is always a more or less jagged and irregular line.

Onward from new moon toward full the horns of the crescent are always turned upward or eastward. When the general line of the terminator becomes a straight line from cusp to cusp, the moon is said to have reached first quarter or quadrature. Onward toward full the terminator will be seen to bend the other way, and in about a week's time it will have merged itself with the moon's limb. The moon is then said to be full. Afterward the phase phenomena recur in the reverse order, with third quarter midway between full and new moon again; the phase of the moon being called gibbous all the way from first quarter to third quarter, except when exactly full.

As we know that the moon is, like the earth, a nonluminous body, and shines only by virtue of the sunlight falling upon it, clearly an entire half of the moon's globe must be perpetually illumined by sunlight. The varying phases then are due simply to that part of the illuminated hemisphere which is turned toward us. New moon is entirely invisible because the sunward hemisphere is turned wholly away from us, while at full moon we see the lunar disk complete because we are on the same side of the moon that the sun is and practically in line with both sun and moon.

If we could visit the moon, we should see the earth in exactly complementary phase. At new moon here we should be enjoying full earth there, and full moon here would be coincident with new or dark earth there. The narrow crescent of new moon here would be the period of gibbous earth there; and it is the reflection of sunlight from this gibbous earth which illuminates the part of the moon but faintly seen at this time, popularly known as the "old moon in the new moon's arms." Its greater visibility at some times than at others is due to greater prevalence of clouded area in the reflecting regions of the earth turned toward the moon, and the higher reflective power of clouds than that possessed by mere land and water.

As the moon goes all the way round the sky every month, the same as the sun does in a year, and travels in nearly the same path, clearly it must also go north and south every month as the sun does. So in midsummer when the sun runs high upon the meridian, we expect to find full moons running low, and likewise in midwinter the full moon always runs high, as almost everyone has sometimes or other noticed.

This eastward or true orbital motion of the moon is responsible for another relation which soon comes to light when we begin to observe the moon; and that is the later hour of rising or setting each night. Our clock time is regulated by the sun, which also is moving eastward about 1° daily, or twice its own breadth. So the moon's eastward gain on the sun amounts to about 12 degrees daily, and one degree being equal to 4 minutes, the retarded time of moonrise or moonset each day amounts to very nearly 50 minutes on the average; though sometimes the delay will be less than a half hour and at other times it will exceed an hour and a quarter. The season of least retardation of rising of the full moon is in the autumn, and so the moon that falls in late September or October is known as the Harvest moon, and the next succeeding full moon is called the Hunter's moon.

Lunation is a term sometimes given to the moon's period from any definite phase round to the same phase again. Its length is the true period of the moon's revolution once around the earth, from the sun all the way round till it overtakes the sun again. The synodic period is another name for lunation, and its true length is 29 and one-half days, or very accurately 29 d. 12 h. 44 m. 2.7 s. as calculated by astronomers with great exactness from many thousand revolutions of the moon. But if we want the true period of the moon round the earth as referred to a star, it is much shorter than this, amounting to only 27 days and nearly one-third. This is called the moon's sidereal period of revolution, because it is the time elapsed while she is traveling eastward from a given star around to coincidence with the same star again.

If we study the moon's path in the sky more critically, we shall find that it does not quite follow the ecliptic, or the sun's path, but that twice each month she deviates from the ecliptic, once to the north and once to the south of it, by roughly ten times her own breadth. More accurately this angle is 5°8'40", an almost invariable quantity, and it is therefore known as an astronomical constant, or the inclination of the moon's orbit to the ecliptic. So the moon's orbit must intersect the ecliptic, and as both are great circles in the sky, the points of intersection are known as the moon's nodes, one ascending and the other descending, and the nodes are 180 degrees apart.

The figure of the moon's orbit is not circular, although it deviates only slightly from that form. But like the paths of all other satellites round their primary planets, and of the planets themselves round the sun, the moon's orbit is also an ellipse. The distance of the moon's center from the earth's center is therefore perpetually changing; the point of nearest approach is called perigee, and that of farthest recession, apogee.

The moon's distance from the earth is easier and simpler to be ascertained than that of any other heavenly body, because it is the nearest. An outline of the method of finding this distance is not difficult to present; and it resembles in every particular the method a surveyor uses to find the distance of some inaccessible point which he cannot measure directly. Up and down a stream, for example, he measures the length of a line, and from each end of it he measures the angle between the other end of the line and the object on the opposite side of the stream whose distance he wishes to find out. Then he applies the science of trigonometry to these three measures, two of angles and one the length of the side or base included between them, and a few minutes' calculation gives the distance of the inaccessible object from either end of the base line.

Now in like manner, to transfer the process to the sky, let the two ends of the base be represented by two astronomical observatories, for example, Greenwich in the northern hemisphere and Cape Town in the southern. The base line is the chord or straight line through the earth connecting the two observatories, and we know the length of this line pretty accurately, because we know the size of the earth. The angles measured are somewhat different from those in the terrestrial example, but the process amounts to the same thing because the astronomers at the two observatories measure the angular distance of the center of the moon from the zenith, each using his own zenith at the same time; and the same science of trigonometry enables them to figure out the length of any side of the triangles involved. The side which belongs to both triangles is the distance from the center of the earth to the center of the moon, and the average of many hundred measures of this gives 238,800 miles, or about ten times the distance round the equator of the earth.

We have said that the orbit in which the moon travels round the earth is practically a circle, but the earth's center is found not at the center of this orbit, but set to one side, or eccentrically, so that the distance spanning the centers of the two bodies is sometimes as small as 221,610 miles at perigee, and 252,970 miles at apogee. The moon's speed in this orbit averages rather more than half a mile every second of time—more accurately 3,350 feet a second, or 2,290 miles per hour.

Once the moon's distance is known, its size or diameter is easy to ascertain. An angular measure is necessary, of course, that of the angle which the disk of the moon fills as seen from the earth. There are many types of astronomical instruments with which this angle can be measured, and its value is something more than half a degree (31' 7"). The moon's actual diameter figures out from this 2,163 miles; and it would therefore require nearly fifty moons merged in one to make a ball the size of the earth.

Still, no other planet has a satellite as large in proportion to its primary as the moon is in relation to the earth. But the materials that compose the moon have less than two-thirds the average density of those that make up the earth, so that eighty-one moons fused together would be necessary to equal the mass or weight of the earth. If we figure out the force of attraction of the moon for bodies on its surface, we find it equals about one-sixth that of the earth. Athletes could perform some astounding feats there—miracles of high jump and hammer-throw.

Our interest in the moon's physical characteristics never wanes. Her nearness to us has always fascinated astronomer and layman alike. Early users of the telescope were readily led into error regarding the general characteristics of the lunar surface; and it is easy to see why they thought the smooth level planes must be seas, and gave them names to that effect which persist to-day, as Mare Crisium, Mare Serenitatis and so on. We may be sure that no water exists on the moon's surface, although some astronomers think that solid water, as ice or snow, may still exist there at a temperature too low for appreciable evaporation.

Perhaps water, seas, and oceans were once there, but their secular dissemination and loss as vapor have gone on through the millions of millions of years till even the moon's atmosphere appears to have vanished completely. At least there is much better evidence of absence of atmosphere on the moon than of its presence—not enough at any rate to equal a thousandth part of the barometric pressure that we have at the earth's surface. Frequent observations of stars passing behind the moon in occultation have satisfied astronomers on this point.

We often say of the brilliant full moon, it is as bright as day. The photometer or instrument for accurate comparison of lights, their amount and intensity, tells a different story. Indeed, if the entire dome of the sky were filled with full moons, we should be receiving only one-eighth of the light the sun gives us, and it would require more than 600,000 average full moons to equal the light radiation of the sun. Heat from the moon, however, is quite different. Early attempts to measure it detected none at all, but with modern instruments there is little trouble in detecting heat from the moon, though measurement of it is not easy.

Much of the moon's heat is sun heat, directly reflected from the moon, as sunlight is, but most of it is due to radiation of solar heat previously absorbed by the materials of the lunar surface. The actual temperature of the moon's surface suffers great variation. A fortnight's perpetual shining of the sun upon the lunar rocks would certainly heat them above the temperature of boiling water, if the moon had an atmosphere to conserve and store this heat; but the entire absence of such an air blanket probably permits the sun's heat to be radiated away nearly as fast as it is received, leaving the temperature at the surface always very low.

What physical influences the moon really has upon the earth must be very slight, barring the tides. But there is little hope of getting people generally to take that view, because the moon appears to be the planet of the people, and opinion that the moon controls the weather, for instance, amounts with them to practical certainty. More than likely all these notions are but legitimate survivals of super-stition and astrology. In addition to the tides, our magnetic observatories reveal slight disturbances with the swinging of the moon from apogee to perigee and back; but long series of weather observations have been faithfully interrogated, with negative or contradictory results. If one believes that the moon's changes affect the weather, it is easy to remember coincidences, and pass over the many times when no change has taken place. The moon changes pretty frequently any-how. As Young well puts it: "A change of the moon necessarily occurs about once a week…. *All* changes, of the weather for instance, must therefore occur within three and three-fourth days of a change of the moon, and fifty per cent of them ought to occur within forty-six hours of a change, even if there were no causal connection whatever."

When we turn to the strongly diversified surface of the moon itself, we find much to rivet the attention, even with slender optical aid. Everyone wants to know how near the telescope, the biggest possible telescope, brings the moon to us. That will depend on many things, first of all on the magnifying power of the eyepiece employed on the telescope, and eyepieces are changed on telescopes just as they are on microscopes, though not for the same reasons. The theoretical limit of the power of a telescope is usually considered as 100 for each inch of diameter or ap-erture of the object glass.

A 40-inch telescope, as that of the Yerkes Observatory, the largest refracting telescope in existence, should bear a magnifying power not to exceed 4,000. But this limit is practically never reached, one-half of it or fifty to the inch of aperture being a good working limit of power, even under exceptional conditions of steadi-ness of atmosphere. If we reduce the effective distance of the moon from 240,000 miles to 100 miles, that is about the utmost that can be expected. But even at that distance we can make out only landscape details, nothing whatever like buildings or the works of intelligence.

The larger relations of light and shade, so obvious to the naked eye on the moon, vanish on looking at it with the telescope, but we are at once captivated by the novel character of the surface and the seemingly great variety of detail that is clearly visible. As soon as the new moon comes out in the west, one may begin to gaze with interest and watch the terminator or sunrise line gradually steal over the roughened surface, bringing new and striking craters into view each night. Around the time of quarter moon, or a little past it, is one of the best times for telescopic views of the moon, because the huge craters, Tycho and Copernicus, are then in fine illumination. Close to the phase of full moon is never a good time, because there are no shadows of the rough surface then, and its entire structure seems to be quite flat and uninteresting, except for the streaks or rills which radiate from Tycho in every direction, and are the only lunar features that are best seen near full.

In a broad, general way, the moon's surface, if compared with the earth's, differs in having no water. Our extensive oceans are replaced there by smooth, level plains which were at first thought to be seas and so named. There are ten or twelve of them in all. Then we find mountain ranges, so numerous on the earth, relatively few on the moon. Those that exist are named, in part, for terrestrial mountain ranges, as the Alps, Caucasus, and the Apennines.

But the nearly circular crater, a relatively rare formation on the earth, is seen dotted all over the moon in every size, from a fraction of a mile in diameter up to sixty, seventy, and in extreme cases a hundred miles. No mere description of plains and mountains and craters affords an adequate idea of the moon's surface as it actually is; a telescopic view is necessary, or some of the modern photographs which give an even better notion of the moon than any telescopic view. Many of the lunar craters are without doubt volcanic in origin, others seem to be ruins of molten lakes. Many thousands of the smaller ones appear as if formed by a violent pelting of the surface when semi-plastic, perhaps by enormous showers of meteoric matter. More than 30,000 craters cover the half of the lunar surface visible from the earth, and hundreds of them are named for philosophers and astronomers.

Measurement of the height of lunar mountains has been made in numerous instances, especially when their shadows fall on plains or surfaces that are nearly level, so that the length of the shadow can be measured. In general, the height of lunar peaks is greater than that of terrestrial peaks, owing probably to the lesser surface gravity on the moon. About forty lunar peaks are higher than Mont Blanc.

Most astronomers regard it as certain that no changes ever take place on the moon; probably no very conspicuous changes ever do. Some, however, have made out a fair case for comparatively recent changes in surface detail. Extreme caution is necessary in drawing conclusions, because the varying changes of illumination from one phase to another are themselves sufficient to cause the appearance of change. At intervals of a double lunation, equal to fifty-nine days, one and one-half hours, the terminator goes very nearly through the same objects, so that the circumstances of illumination are comparable. In Mare Serenitatis the little crater named Linné was announced to have disappeared about a half century ago; subsequently it became visible again and other minor changes were reported, perhaps due to falling in of the walls of the crater.

If one were to visit the moon, he must needs take air and water along with him, as well as other sustenance. No atmosphere means no diffused light; we could see nothing unless the sun's direct rays were shining upon it. Anyone stepping into the shadow of a lunar crag would become wholly invisible. No sound, however loud, could be heard; sound in fact would become impossible. A rock might roll down the wall of a lunar crater, but there would be no noise; though we should know what had happened by the tremor produced. So slight is gravity there that a good ball player might bat a baseball half a mile or more. Looking upward, all the stars would be appreciably brighter than here, and visible perpetually in the daytime as well as at night.

If one were to go to the opposite side of the moon, he would lose sight of the earth until he came back to the side which is always turned toward the earth. Even then the earth would never rise and set at any given place, as the moon does to us, but would remain all the time at about the same height above the lunar horizon. The earth would go through all the phases that the moon shows to us here, full earth occurring there when it is new moon here. Our globe would appear to be nearly four times broader than the moon seems to us. Its white polar caps of ice and snow, its dark oceans, and the vast cloud areas would be very conspicuous. Faint stars, the zodiacal light, and the filmy solar corona would be visible, probably even close up to the sun's edge; but although his rays might shine upon the lunar rocks without intermission for a fortnight, probably they would still be too cold to touch with safety. On the side of the moon turned away from the sun, the temperature of the moon's surface would fall to that of space, or many hundred degrees below zero.

CHAPTER XXIX
ECLIPSES OF THE MOON

Of all the weird happenings of the nighttime sky, eclipses of the moon are the most impressive. Rarely is there a year without one. What is the cause? Simply the earth getting in between sun and moon, and thereby shutting off the sunlight which at all other times enables us to see the moon. As the earth is a dark body it must cast a black shadow on the side away from the sun, and it is the moon's passing into this shadow or some part of it that causes a lunar eclipse.

Sun and earth being so different in size, the earth's shadow must stretch away from it into space, growing smaller and smaller, until at length it comes to an end—the apex of a cone 857,000 miles long. If we cut off this shadow at the moon's distance from the earth, we find it about 6,000 miles in diameter at that point; and this accounts for the fact that the curvature on the side of the moon, when the eclipse is coming on and where it is dropping into the shadow, is always much less rapid than the curvature of the moon's own disk is.

When an eclipse is approaching, the eastern limb will be duskily darkened for half an hour or more, because the moon must first pass through the outer penumbra, or half-shadow which everywhere surrounds the true shadow itself. If the moon hits only the upper or lower part of the shadow, the eclipse will be only partial, and during the progress of the eclipse it will seem as if the uneclipsed part had swung or twisted around in the sky, from the western limb of the moon to the eastern. But when the moon passes through the middle regions of the shadow, the eclipse is always total, and direct sunlight is wholly cut off from every part of the moon's face, for a greater or less length of time, according to the part of the shadow through which it passes. When passing centrally through the shadow, the total eclipse will last about two hours, as the moon's diameter is about one-third of the breadth of the shadow; and the eclipse will be partial about two hours longer, an hour at beginning and an hour at the end, because the moon moves over her own breadth in about an hour.

While the moon is wholly immersed in the shadow, her body is nevertheless visible, as a dull tarnished copper disk; and this is caused by the reddish sunlight which grazes the earth all around and is refracted or bent by our atmosphere into the shadow itself. If this belt or ring of terrestrial atmosphere happens to be everywhere filled with dense clouds, as was the case in 1886, even the familiar copper moon of a total lunar eclipse disappears completely in the black sky.

Quite different from a solar eclipse, all the phases of a lunar eclipse are visible at the same time on the earth wherever the moon is above the horizon. Eclipses of the moon are therefore seen with great frequency at any given place as compared with solar eclipses, which are restricted to relatively narrow areas of the earth's surface. Nor are lunar eclipses of very much significance to the astronomer, mainly because of the slowness and indefiniteness of the phenomena. It is a good time to observe occultations of faint stars at the moon's edge or limb, and several such programs have been carried out by cooperation of observatories in widely separate regions of the world: the object being improvement in our knowledge of the distance of the moon, and in the accuracy of the mathematical tables of her motion. Search by photography for a possible satellite, or moon of the moon, has been made on several occasions, though without success.

A lunar eclipse was first observed and photographed from an aeroplane, May 2, 1920. At the request of the writer, two aviators of the United States navy ascended to a height of 15,000 feet above Rockaway, and secured many advantages accruing from great elevation in viewing a celestial phenomenon of this character.

CHAPTER XXX
TOTAL ECLIPSES OF THE SUN

Primitive peoples indulged in every variety of explanation of mysterious happenings in the sky. To the Chinese and all through India, a total eclipse of the sun is caused by "a certain dragon with very black claws," who, except for their frightening him away by every conceivable sort of hideous noise, would most certainly "eat up the sun." The eclipse always goes off, the sun has never been eaten yet. Can you convince a Chinaman that Rahu, the Dragon, wouldn't have eaten up the sun, if his unearthly din hadn't frightened him away?

In Japan the eclipse drops poison from the sky into wells, so the Japanese cover them up. Fontenelle relates that in the middle of the seventeenth century a multitude of people shut themselves up in cellars in Paris during a total eclipse.

In the Shu-king, an ancient Chinese work, occurs the earliest record of a total eclipse of the sun, in the year B. C. 2158. The Nineveh eclipse of B. C. 763 is perhaps the first of the ancient eclipses of which we possess a really clear description on the Assyrian eponym tablets in the British Museum. It is the eclipse possibly referred to in the Book of Amos, viii.

But of all the ancient eclipses none perhaps exceeds in interest the famous eclipse of Thales, B. C. 585, May 28. It is the first eclipse to have been predicted, probably by means of the saros, or 18-year period of eclipses, which is useful as an approximate method even at the present day. But the accident of a war between the Lydians and the Medes has added greatly to the historic interest, because the combatants were so terrified by the sudden turning of day into night that they at once concluded a peace cemented by two marriages.

Very many of the ancient eclipses have been of great use to the historian in verifying dates, and mathematical astronomers have employed them in correcting the lunar tables, or intricate mathematical data by which the motion of the moon is predicted.

Coming down to the middle of the sixth century, we find the first eclipse recorded in England, in the "Saxon Chronicle," A. D. 538. During the epoch of the Arabian Nights several eclipses were witnessed at Bagdad, A. D. 829 to 928, and many a century later by Ibu-Jounis, court astronomer of Hakem, the Caliph of Egypt. Nothing is more interesting than to search the quaint records of these ancient eclipses. One occurring in 1560, when Tycho Brahe was but fourteen, had much to do with turning his permanent interest toward mathematics and astronomy. The eclipse of 1612 was the first "seen through a tube," the telescope having been invented only a few years before. "Paradise Lost" was completed about 1665, and the censorship was still in existence; and it is matter of record that the oft-quoted passage,

> "As when the Sun, new risen,
> Looks through the horizontal misty air,
> Shorn of his beams; or from behind the Moon,
> In dim eclipse, disastrous twilight sheds
> On half the nations, and with fear of change
> Perplexes monarchs."

P. L., i. 594

was strongly urged as sufficient reason for suppressing the entire epic.

London was favored with the outflashing corona, May 3, 1715, and a pamphlet was issued in prediction, entitled "The Black Day, or a Prospect of Doomsday."

The first American eclipse expedition was on occasion of the totality of Oct. 27, 1780, sent out by Harvard College and the American Academy of Arts and Sciences under Professor Samuel Williams to Penobscot. There was a fine total eclipse from Albany to Boston on June 16, 1806, and many important observations of it were made in this country.

But it was not till the European eclipse of 1842 that research got fully under way, because the germ of the new astronomy, particularly as applied to the sun, had begun its development; and the significance of the corona was obvious, if it could be proved a true appendage of the sun. Photography had not long been discovered, and the corona of 1851 was the first to be automatically registered on a daguerreotype. In 1860 it was proved that prominences and corona both belong to the sun and not to the moon.

The great Indian eclipse of 1868 brought the important discovery that the prominences can be observed at any time without an eclipse by means of the spectroscope. In 1869 bright lines were found in the spectrum of the corona, one line in the green indicating the presence of an element not then known on the earth and hence called coronium. In 1870 the reversing layer or stratum of the sun was discovered. In 1878 a vast ecliptic extension of the streams of the corona many millions of miles both east and west of the sun was first seen. This is now known to be the type of corona characteristic of minimum spots on the sun. In 1882 the spectrum of the corona was first photographed and in 1889 excellent detail photographs of the corona were taken. In 1893 it was shown that the corona quite certainly rotates bodily with the sun. In 1896 actual spectrum photographs of the reversing layer established its existence beyond doubt—"flash spectrum" it is often called. In 1898 the long ecliptic streamers of the corona were successfully photographed for the first time. In 1900 the depth of the reversing layer was found to average 500 miles, the heat of the corona was first measured by the bolometer, and many observations showed that the coronal streamers, in part at least, partake of the nature of electric discharges.

All subsequent total eclipses have been carefully observed, in whatever part of the world they may happen, and each has added new results of significance to our theories of the corona and its relation to the radiant energy of the sun. In very recent eclipses the cinematograph has been brought into action as an efficient adjunct of observation; in 1914 the first successful "movie" of the eclipse was secured in Sweden, and in 1918 Frost of the Yerkes Observatory first applied the cinematograph to registry of the "flash spectrum," and Stebbins tested out his photo-electric cell on the corona, making the brightness 0.5 that of the full moon. In 1914 (Russia) and again in 1919 (on the Atlantic) the obvious advantages of the aeroplane in ecliptic observation and photography were sought by the writer, though unsuccessfully. The photographic tests, however, conducted in preparation for these expeditions proved the entire practicability of securing eclipse results of much value, independently of clouds below.

Eclipses in the near future will be total in Australia about six minutes on September 21, 1922; in California and Mexico about four minutes on September 10, 1923; and along a line from Toronto to Nantucket about two minutes on the morning of January 24, 1925.

To all spectators, savage or civilized, scientist or layman, a total eclipse is wonderful and impressive. Langley said: "The spectacle is one of which, though the man of science may prosaically state the facts, perhaps only the poet could render the impression." Very gradually the moon steals its way across the face of the sun, the lessened light is hardly noticed. If one is near a tree through whose foliage the sunlight filters, an extraordinary sight is seen; the ground all about is covered with luminous crescents, instead of the overlapping disks which were there before the eclipse came on; in both cases they are images of the disk of the sun at the time, and the narrowing crescents will be watched with interest as totality approaches. Then the shadow bands may be seen flitting across the landscape, like "visible wind." They are probably related to our atmosphere and the very slender crescent from which true sunlight still comes.

Then for a few seconds the moon's actual shadow may be caught in its approach, very suddenly the darkness steals over the landscape and—totality is on. How lucky if there are no clouds! Every eye is riveted on "the incomparable corona, a silvery, soft, unearthly light, with radiant streamers, stretching at times millions of uncomprehended miles into space, while the rosy flaming protuberances skirt the black rim of the moon in ethereal splendor."

Then it is now or never with observer and photographer. Months of diligent preparations at home followed by weeks of tedious journey abroad, with days of strenuous preparation and rehearsals at the station—all go for naught unless the whole is tuned up to perfect operation the instant totality begins. It may last but a minute, or even less; in 1937, however, total eclipse will last 7 minutes 20 seconds, the longest ever observed, and within half a minute of the longest possible. All is over as suddenly as it came on. The first thing is to complete records, develop plates, and see if everything worked perfectly.

There is great utility back of all eclipse research, on account of its wide bearing on meteorology and terrestrial physics, and possibly the direct use of solar energy for industrial purposes. With this purpose in view the astronomer devotes himself unsparingly to the acquisition of every possible fact about the sun and his corona.

Considering the earth as a whole, the number of total eclipses will average nearly seventy to the century. But at any given place, one may count himself very fortunate if he sees a single total eclipse, although he may see several partial ones without going from home. Then, too, there are annular or ring eclipses, averaging seven in eight years. But had one been born in Boston or New York in the latter part of the eighteenth century, he might have lived through the entire nineteenth century and a long way into the twentieth without seeing more than one total eclipse of the sun. In London in 1715 no total eclipse had been visible for six centuries. However, taking general averages, and recalling the comparatively narrow belt of total eclipse, every part of the earth is likely to come within range of the moon's shadow once in about three and a half centuries.

The longest total eclipses always occur near the equator; this is because an observer on the equator is carried eastward by the earth's rotation at a velocity of about 1,000 miles per hour, so that he remains longer in the moon's shadow which is passing over him in the same direction with a velocity about twice as great.

The general circumstances of total eclipses are readily foretold by means of the ancient Chaldean period of eclipses known as the saros. It is 18 years and 10 or 11 days in length (according to the number of leap years intervening). In one complete saros, forty-one solar eclipses will generally happen, but only about one-fourth of them will be total. The saros is a period at the end of which the centers of sun and moon return very nearly to their relative positions at the beginning of the cycle. So, in general, the eclipse of any year will be a repetition of one which took place 18 years before, and another very similar in circumstances will happen 18 years in the future. Three periods of the saros, or 54 years and 1 month, will usually bring about a return of any given eclipse to any particular part of the earth, so far as longitude is concerned, though the returning track will lie about 600 miles to the north or south of the one 54 years earlier.

Paths of total eclipses frequently intersect, if large areas like an entire country are considered; Spain, for instance, where total eclipses have occurred in 1842, 1860, 1870, 1900 and 1905. Besides crossing Spain, the tracks of totality on May 28, 1900, and August 30, 1905, were unique in intersecting exactly over a large city — Tripoli in Barbary, on both of which occasions the writer's expeditions to that city were rewarded with perfect observing conditions in that now Italian province on the edge of the great desert.

Kepler was the first astronomer to calculate eclipses with some approach to scientific form, as exemplified in his Rudolphine Tables. His method was of course geometrical. But La Grange, who applied the methods of more refined analysis to the problem, was the first to develop a method by which an eclipse and all its circumstances could be accurately predicted for any part of the earth. To many minds, the prediction of an eclipse affords the best illustration of the superior knowledge of the astronomer: it seems little short of the marvelous. But recalling that the motion of the moon follows the law of gravitation, and that its position in the sky is predictable for years in advance with a high degree of precision, it will readily be seen how the arrival of the moon's shadow, and hence the total eclipses of the sun, can be foretold for any place over which the shadow passes.

All these data derived by the mathematician are known as the elements of the eclipse, and they are prepared many years in advance and published in the nautical almanacs and astronomical ephemerides issued by the leading nations. Buchanan's "Treatise on Eclipses" will supply all the technical information regarding the prediction of eclipses that anyone desirous of inquiring into this phase of the problem may desire.

So important are total eclipses in the scheme of modern solar research, and so necessary are clear skies in order that expeditions may be favored with success, that every effort is now made to ascertain the weather chances at particular stations along the line of eclipse many years in advance. This method of securing preliminary cloud observations for a series of years has proved especially useful for the eclipses of 1893, 1896, 1900, and 1918; and had it been employed in Russia for totality of 1914, many well-equipped expeditions might have been spared disaster. The California and Mexico totality of 1923 does not require this forethought, as the regions visited are quite likely to be free from cloud; but observations are now in process of accumulation for the total eclipse of 1925. The out-look for clear skies on that occasion, the total eclipse nearest New York for more than a century, is not very promising. The path of totality passes over Marquette, Michigan, Rochester and Poughkeepsie, New York, Newport, Rhode Island, and Nantucket about nine in the morning.

Everyone who saw it will remember the last total eclipse in this part of the world—on June 8, 1918, visible from Oregon to Florida. Many will recall the last total eclipse that was visible before that in the eastern part of the United States, on May 28, 1900, visible in a narrow path from New Orleans to Norfolk. One's father or grandfather will perhaps remember the total eclipse of July 29, 1878, which passed over the United States from Pike's Peak to Texas (it was the writer's maiden eclipse), and another on August 7, 1869, which passed southeasterly over Iowa and Kentucky. On all these occasions the paths of total eclipse were dotted with numerous observing parties, many of them equipped with elaborate apparatus for studying and photographing the solar corona and prominences, together with a multitude of other phenomena which are seen only when total eclipses take place.

Looking forward rather than backward, a striking series, or family, of eclipses happens in the future: it is the series of May, 1901 and 1919, recurring again on June 8, 1937 (over the Pacific Ocean), June 20, 1955 (through India, Siam, and Luzon), and June 30, 1973 (visible in Sahara, Abyssinia, and Somali). Already in 1919 this totality was 6 minutes 50 seconds in duration; in 1937, as already mentioned, it will be 7 minutes 20 seconds, and at the subsequent returns even longer yet, approaching the estimated maximum of 7 minutes 58 seconds which has never been observed. This remarkable series of total eclipses is longer in duration than any others during a thousand years. Its next subsequent return is in 1991, occurring with the eclipsed sun practically at noon in the zenith of Mount Popocatepetl in Mexico.

Whatever may be the progress of solar research during the intervening years, it is impossible to imagine the alert astronomer of that remote day without incentive for further investigation of the sun's corona, in which are concealed no doubt many secrets of the sun's evolution from nebula to star.

CHAPTER XXXI

THE SOLAR CORONA

"And what is the sun's corona?" mildly asked a college professor of a student who might better have answered "Not prepared."

"I did know, Professor, but I have forgotten," was his reply.

"What an incalculable loss to science," returned the professor with a twinkle. "The only man who ever knew what the sun's corona is, and he has forgotten!"

Only in part has the mystery of the corona been cleared by the research of the present day. Our knowledge proceeds but slowly, because the corona has never been seen except during total eclipses of the sun; and astronomers, as a matter of fact, have never had a fair chance at it. Two total eclipses happen on the average of every three years; their average duration is only two or three minutes; totality can be seen only in a narrow path about a hundred miles wide, though it may be several thousand miles long; there is usually about equal chance of cloud with clear skies; and fully three-fourths of the totality areas of the globe are unavailable because covered by water. So that even if we imagine the tracks of eclipses quite thickly populated with astronomers and telescopes, at least one every hundred miles, how much solid watching of the corona would this permit? Only a little more than one week's time in a whole century.

The true corona is at least a triple phenomenon and a very complex one. The photographs reveal it much as the eye sees it, with all its complexity of interlacing streamers projected into a flat, or plane, surrounding the disk of the dark moon which hides the true sun completely. But we must keep in mind the fact that the sun is a globe, not a disk, and that the streamers of the corona radiate more or less from all parts of the surface of the solar sphere, much as quills from a porcupine.

From the sun's magnetic poles branch out the polar rays, nearly straight throughout their visible extent. Gradually as the coronal rays originate at points around the solar disk farther and farther removed from the poles, they are more and more curved. Very probably they extend into the equatorial regions, but it is not easy to trace them there because they are projected upon and confused with the filaments having their origin remote from the poles. Then there is the inner equatorial corona, apparently connected intimately with truly solar phenomena, quite as the polar rays are. The third element in the composite is the outer ecliptic corona, for the most part made up of long streamers. This is most fully developed at the time of the fewest spots on the sun. It is traceable much farther against the black sky with the naked eye than by photography. Without any doubt it is a solar appendage and possibly it may merge into the zodiacal light.

Naturally this superb spectacle must have been an amazing sight to the beholders of antiquity who were fortunate enough to see it. Historical references are rare: perhaps the earliest was by Plutarch about A. D. 100, who wrote of it, "A radiance shone round the rim, and would not suffer darkness to become deep and intense." Philostratus a century later mentions the death of the emperor Domitian at Ephesus as "announced" by a total eclipse.

Kepler thought the corona was evidence of a lunar atmosphere; indeed, it was not until the middle of the 19th century that its lack of relation to the moon was finally demonstrated. Later observers, Wyberd in 1652 and Ulloa, got the impression that the corona turned round the disk catherine-wheel fashion, "like an ignited wheel in fireworks, turning on its center." But no later observer has reported anything of the sort. Quite the contrary, there it stands against the black sky in motionless magnificence a colorless pearly mass of wisps and streamers for the most part nebulous and ill-defined, fading out very irregularly into the black sky beyond, but with a complex interlacing of filaments, sometimes very sharply defined near the solar poles. It defies the skill of artist and draughtsman to sketch it before it is gone.

Photograph it? Yes, but there are troubles. Of course the camera work is superior to sketches by hand. As Langley used to say, "The camera has no nerves, and what it sets down we may rely on." Foremost among the photographic difficulties is the wide variation in intensity of the coronal light in different regions of the corona. If a plate is exposed long enough to get the outer corona, the exceeding brightness of the inner corona overexposes and burns out that part of the plate or film. If the exposure is short, we get certain regions of the inner corona excellently, but the outer regions are a blank because they can be caught only by a long exposure.

So the only way is to take a series of pictures with a wide range of exposures, and then by careful and artistic handwork, combine them all into a single drawing. Wesley of London has succeeded eminently in work of this character, and his drawings of the sun's corona, visible at total eclipses from 1871 onward, in possession of the Royal Astronomical Society, are the finest in existence. They give a vastly better idea of the corona, as the eye sees it, than any single photograph possibly can.

The early observers apparently never thought of the corona as being connected with the sun. It was a halo merely, and so drawn. Its real structure was neither known, depicted, or investigated. Sketches were structureless, as any aureola formed by stray sunlight grazing the moon might naturally be. That the rays are curved and far from radial round the sun was shown for the first time in the sketches of 1842, and in 1860 Sir Francis Galton observed that the long arms or streamers "do not radiate strictly from the center."

The inner corona had first been recorded photographically on a daguerreotype plate during the eclipse of 1851, but the lens belonged to a heliometer, and was of course uncorrected for the photographic rays. The wet collodion plates of the eclipse of 1860, by De la Rue, proved that not only the prominences but the corona were truly solar, because his series of technically perfect pictures revealed the steady and unchanged character of these phenomena while the moon's disk was passing over them as totality progressed. And at the eclipse of 1869, Young put the solar theory of the corona beyond the shadow of any further doubt by examination of its light with the spectroscope and discovering a green line in the spectrum due to incandescent vapor of a substance not then identified with anything terrestrial, and therefore called coronium.

The total brilliance of the corona was very differently estimated by the earlier observers, though pretty carefully measured at later eclipses. The standard full moon is used for reference, and at one eclipse the corona falls short of, while at another it will exceed the full moon in brightness. Variations in brilliancy are quite marked: at one eclipse it was nearly four times as bright as the full moon. Much evidence has already accumulated on this question; but whether the observed variations are real, or due mainly to the varying relative sizes of sun and moon at different eclipses, is not yet known. The coronal light is largely bluish in tint, and this is the region of the spectrum most powerfully absorbed by our atmosphere. Eclipses are observed by different expeditions located at stations where the eclipsed sun stands at very different altitudes above the horizon; besides this the localities of observation are at varied elevations above sea level; so that the varying amount of absorption of the coronal light renders the problem one of much difficulty.

The long ecliptic streamers of the corona were first seen by Newcomb and Langley during the totality of 1878. On one side of the sun there was a stupendous extension of at least twelve solar diameters, or nearly 11 millions of miles. Langley observed from the summit of Pike's Peak, over 14,000 feet high, and was sure that he was witnessing a "real phenomenon heretofore undescribed." The vast advantage of elevation was apparent also from the fact that he held the corona for more than four minutes after true totality had ended. These streamers are characteristic of the epoch of minimum spots on the sun, as Ranyard first suggested. It was found that this type of corona had been recorded also in 1867; and it has reappeared in 1889, 1900 and 1911, and will doubtless be visible again in 1922.

How rapidly the streamers of the corona vary is not known. Occasionally an observer reports having seen the filaments vibrate rapidly as in the aurora borealis, but this is not verified by others who saw the same corona perfectly unmoving. Comparisons of photographs taken at widely separate stations during the same eclipse have shown that at least the corona remained stationary for hours at a time. Whether it may be unchanged at the end of a day, or a week, or a month, is not known; because no two total eclipses can ever happen nearer each other than within an interval of 173 days, or one-half of the eclipse year. And usually the interval between total eclipses is twice or three times this period.

Theories of what the solar corona may be are very numerous. The extreme inner corona is perhaps in part a sort of gaseous atmosphere of the sun, due to matter ejected from the sun, and kept in motion by forces of ejection, gravity, and repulsion of some sort. Meteoric matter is likely concerned in it, and Huggins suggested the débris of disintegrating comets. Schuster was in agreement with Huggins that the brighter filaments of the corona might be due to electric discharges, but it seems very unlikely that any single hypothesis can completely account for the intricate tracery of so complex a phenomenon.

Solar Corona and Prominences. Photographed during a total eclipse of the sun, June 8, 1918. (Courtesy, American Museum of Natural History.)

Elaborate spectroscopic programs have been carried out at recent eclipses, affording evidence that certain regions are due to incandescent matter of lower temperature than the sun's surface. A small part of the light of the corona is sunlight reflected from dark particles possibly meteoric, but more likely dust particles or fog of some sort. This accounts for the weakened solar spectrum with Fraunhofer absorption lines, and this part of the light is polarized.

Venus, Showing Crescent Phase of the Planet. Venus is the earth's nearest neighbor on the side toward the sun.

(Photo, Yerkes Observatory.)

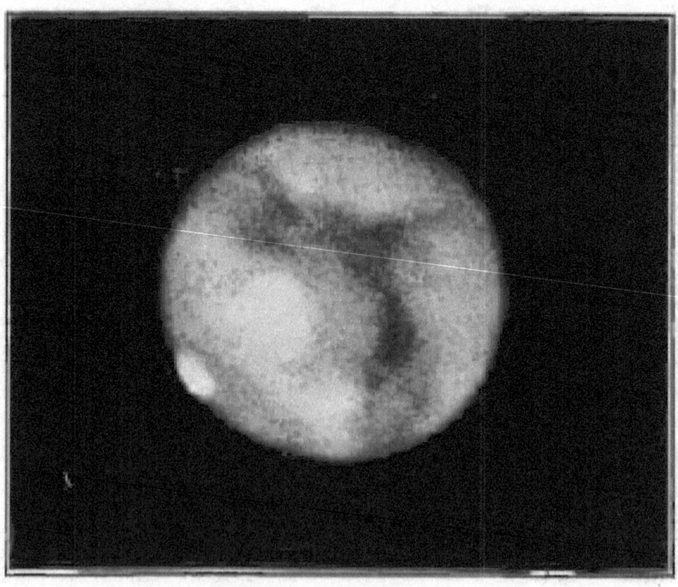

Mars, the Planet Next Beyond the Earth. The photograph shows one of the white polar caps. The caps are thought to be snow or ice and may indicate the existence of atmosphere.

(Photo, Yerkes Observatory.)

Many have been the attempts to see, or photograph, the corona without an eclipse. None of them has, however, succeeded as yet. Huggins got very promising results nearly forty years ago, and success was thought to have been reached; but subsequent experiments on the Riffelberg in 1884 and later convinced him that his results related only to a spurious corona. In 1887 the writer made an unsuccessful attempt to visualize the corona from the summit of Fujiyama, and Hale tried both optical and photographic methods on Pike's Peak in 1893 without success. He devised later a promising method by which the heat of the corona in different regions can be measured by the bolometer, and an outline corona afterward sketched from these results.

Still another method of attacking the problem occurred to the writer in 1919, which has not yet been carried out. It would take advantage of recent advances in aeronautics, and contemplates an artificial eclipse in the upper air by means of a black spherical balloon. This would be sent up to an altitude of perhaps 40,000 feet, where it would partake of the motion of the air current in which it came to equilibrium. Then a snapshot camera would be mounted on an aeroplane, in which the aviator would ascend to such a height that the balloon just covered the sun, as the moon does in a total eclipse. With the center of the balloon in line with the sun's center, he would photograph the regions of the sky immediately surrounding the sun, against which the corona is projected. As the entire apparatus would be above more than an entire half of the earth's atmosphere, the experiment would be well worth the attempt, as pretty much everything else has been tried and found wanting. Needless to say, the importance of seeing the corona at regular intervals whenever desired, without waiting for eclipses of the sun, remains as insistent as ever.

CHAPTER XXXII
THE RUDDY PLANET

Mars is a planet next in order beyond the earth, and its distance from the sun averages 141½ million miles. It has a relatively rapid motion among the stars, its color is reddish, and, when nearest to us, it is perhaps the most conspicuous object in the sky.

Mars appeared to the ancients just as it does to us to-day. Aristotle recorded an observation of Mars, 356 B. C., when the moon passed over the planet, or occulted it, as our expression is. Galileo made the first observations of Mars with a telescope in 1610, and his little instrument was powerful enough to enable him to discover that the planet had phases, though it did not pass through all the phases that Mercury and Venus do. This was obvious from the fact that Mars is always at a greater distance from the sun than we are, and the phase can only be gibbous, or about like the moon when midway between full and quarter.

Many observers in the seventeenth century followed up the planet with such feeble optical power as the telescopes of that epoch provided: Fontana (who made the first sketch), Riccioli and Bianchini in Italy, Cassini in France, Huygens in Holland, and later Sir William Herschel in England.

It was Cassini who first made out the whitish spots or polar caps of Mars in 1666, but not until after Huygens had noted the fact that Mars turned round on an axis in a period but little longer than the earth's. Cassini followed it up later with a more accurate value; and observations in our own day, when combined with these early ones, enable us to say that the Martian day is equal to 24 hours 37 minutes 22.67 seconds, accurate probably to the hundredth part of a second.

When we know that a planet turns round on an axis, we know that it has a day. When we know the direction of the axis in space or in relation to the plane of its path round the sun, we know that it has seasons: we can tell their length and when they begin and end. It did not take many years of observation to prove that the axis round which Mars turns is tilted to the plane of its path round the sun by an angle practically the same as that at which the earth's axis is tilted. So there is the immediate inference that on Mars the order and perhaps the character of the seasons is much the same as here on the earth.

At least two things, however, tend to modify them. First, the year of Mars is not 365 days like ours, but 687 days. Each of the four seasons on Mars, therefore, is proportionally longer than our seasons are. Then comes the question of atmosphere—how much of an atmosphere does Mars really possess in proportion to ours, and how would its lesser amount modify the blending of the seasons into one another?

All discussion of Mars and the problems of existence of life upon that planet hinge upon the character and extent of Martian atmosphere. The planet seems never to be covered, as the earth usually is, with extensive areas of cloud which to an observer in space would completely mask its oceans and continents. Nearly all the time Mars in his equatorial and temperate zones is quite clear of clouds. A few whitish spots are occasionally seen to change their form and position in both northern and southern latitudes, and they vary with the progress of the day on Mars, as clouds naturally would. But Schiaparelli, perhaps the best of all observers, thought them to be not low-lying clouds of the nimbus type that would produce rains, but rather a veil of fog, or perhaps a temporary condensation of vapor, as dew or hoar frost. But the strongest argument for an atmosphere is based on the temporary darkening or obscuration of well known and permanent markings on the surface of Mars. These are more or less frequently observed and clouds afford the best explanation of their occurrence.

So much for evidence supplied by the telescope alone. When, however, we employ the spectroscope in conjunction with the telescope, another sort of evidence is at hand. Several astronomers have reached the conclusion that watery vapor exists in the atmosphere of Mars, while other astronomers equipped with equal or superior apparatus, and under equally favorable or even better conditions, have reached the remarkable conclusion that the spectra of Mars and the moon are identical in every particular. From this we should be led to infer that Mars has perhaps no more atmosphere than the moon has, that is to say, none whatever that present instruments and methods of investigation have enabled us to detect.

What then, shall we conclude? Simply that the atmosphere of Mars is neither very dense nor extensive. Probably its lower strata close to the planet's surface are about as dense as the earth's atmosphere is at the summits of our highest mountains.

This conclusion is not unwelcome, if we keep a few fundamental facts in clear and constant view. Mars is a planet of intermediate size between the earth and the moon: twice the moon's diameter (2,160 miles) very nearly equals the diameter of Mars (4,200 miles), and twice the diameter of Mars does not greatly exceed the earth's diameter (7,920 miles). As to the weights or masses of these bodies, Mars is about one-ninth, and the moon one-eightieth of the earth. The atmospheric envelope of the earth is abundant, the moon has none as far as we can ascertain; so it seems safe to infer that Mars has an atmosphere of slight density: not dense enough to be detected by spectroscopic methods, but yet dense enough to enable us to explain the varying telescopic phenomena of the planet's disk which we should not know how to account for, if there were no atmosphere whatever. One astronomer has, indeed, gone so far as to calculate that in comparison with our planet Mars is entitled to one-twentieth as much atmosphere as we have, and that the mercurial barometer at "sea level" would run about five and a half inches, as against thirty inches on the earth.

In general, then, the climate of Mars is probably very much like that of a clear season on a very high terrestrial table land or mountain—a climate of wide extremes, with great changes of temperature from day to night. The inequality of Martian seasons is such that in his northern hemisphere the winter lasts 381 days and the summer only 306 days.

Now, the polar caps of Mars, which are reasonably assumed to be due to snow or hoar frost, attain their maximum three or four months after the winter solstice, and their minimum about the same length of time after the summer solstice. This lagging should be interpreted as an argument for a Martian atmosphere with heat-storing qualities, similar to that possessed by the earth.

Upon this characteristic, indeed, depends the climate at the surface of Mars: whether it is at all similar to our own, and whether fluid water is a possibility on Mars or not. While the cosmic relations of the planet in its orbit are quite the same as ours, nevertheless the greater distance of Mars diminishes his supply of direct solar heat to about half what we receive. On the other hand, his distance from the sun during his year of motion around it varies much more widely than ours, so that he receives when nearest the sun about one-half more of solar heat than he does when farthest away.

Southern summers on Mars, therefore, must be much hotter, and southern winters colder than the corresponding seasons of his northern hemisphere. Indeed, the length of the southern summer, nearly twice that of the terrestrial season, sometimes amply suffices to melt all the polar ice and snow, as in October, 1894, when the southern polar cap of Mars dwindled rapidly and finally vanished completely.

Very interesting in this connection are the researches of Stoney on the general conditions affecting planetary atmospheres and their composition. According to the kinetic theory, if the molecules of gases which are continually in motion travel outward from the center of a planet, as they frequently must, and with velocities surpassing the limit that a planet's gravity is capable of controlling, these molecules will effect a permanent escape from the planet, and travel through space in orbits of their own.

So the moon is wholly without atmosphere because the moon's gravity is not powerful enough to retain the molecules of its component gases. So also the earth's atmosphere contains no helium or free hydrogen. So, too, Mars is possessed of insufficient force of gravity to retain water vapor, and the Martian atmosphere may therefore consist mainly of nitrogen, argon, and carbon dioxide.

As everyone knows, the axis of the earth if extended to the northern heavens would pass very near the north polar star, which on that account is known as Polaris. In a similar manner the axis of Mars pierces the northern heavens about midway between the two bright stars Alpha Cephei and Alpha Cygni (Deneb). The direction of this axis is pretty accurately known, because the measurement of the polar caps of the planet as they turn round from night to night, year in and year out, has enabled astronomers to assign the inclination of the axis with great precision.

These caps are a brilliant white, and they are generally supposed to be snow and ice. They wax and wane alternately with the seasons on Mars, being largest at the end of the Martian winter and smallest near the end of summer. The existence of the polar caps together with their seasonal fluctuations afford a most convincing argument for the reality of a Martian atmosphere, sufficiently dense to be capable of diffusing and transporting vapor.

The northern cap is centered on the pole almost with geometric exactness, and as far as the 85th parallel of latitude. On the other hand, the south polar cap is centered about 200 miles from the true pole, and this distance has been observed to vary from one season to another. No suggestion has been made to account for this singular variation. On one occasion it stretched down to Martian latitude 70 degrees and was over 1,200 miles in diameter.

Pickering watched the changing conditions of shrinking of the south polar cap in 1892 with a large telescope located in the Andes of Peru. Mars was faithfully followed on every night but one from July 13 to September 9, and the apparent alterations in this cap were very marked, even from night to night. As the snows began to decrease, a long dark line made its appearance near the middle of the cap, and gradually grew until it cut the cap in two. This white polar area (and probably also the northern one in similar fashion) becomes notched on the edge with the progress of its summer season; dark interior spots and fissures form, isolated patches separate from the principal mass, and later seem to dissolve and disappear. Possibly if one were located on Mars and viewing our earth with a big telescope, the seasonal variation of our north and south polar caps might present somewhat similar phenomena. All the recent oppositions of Mars have been critically observed by Pickering from an excellent station in Jamaica.

Quite obviously the fluctuations of the polar caps are the key to the physiographic situation on Mars, and they are made the subject of the closest scrutiny at every recurring opposition of the planet. Several observers, Lowell in particular, record a bluish line or a sort of retreating polar sea, following up the diminishing polar cap as it shrinks with the advance of summer. It is said that no such line is visible during the formation of the polar cap with the approach of winter. All such results of critical observation, just on the limit of visibility, have to be repeated over and over again before they become part of the body of accepted scientific fact. And in many instances the only sure way is to fall back on the photographic record, which all astronomers, whether prejudiced or not, may have the opportunity to examine and draw their individual conclusions.

Already the approaching opposition of 1924, the most favorable since the invention of the telescope, is beginning to attract attention, and preparations are in progress, of new and more powerful instruments, with new and more sensitive photographic processes, by means of which many of the present riddles of Mars may be solved.

CHAPTER XXXIII
THE CANALS OF MARS

Then there are the so-called canals of Mars, about which so much is written and relatively little known. Faint markings which resemble them in character were first drawn in 1840 and later in 1864, but Schiaparelli, the famous Italian astronomer, is probably their original discoverer, when Mars was at its least distance from the earth in 1877. He made the first accurate detailed map of Mars at this time, and most of the important or more conspicuous canals (*canali*, he called them in Italian, that is, channels merely, without any reference whatever to their being watercourses) were accurately charted by him.

At all the subsequent close approaches of Mars, the canals have been critically studied by a wide range of astronomical observers, and their conclusions as to the nature and visibility of the canals have been equally wide and varied. The most favorable oppositions have occurred in 1892 and 1894, also in 1907 and 1909. On these occasions a close minimum distance of Mars was reached, that is, about 35 millions of miles; but in 1924 the planet makes the closest approach in a period of nearly a thousand years. Its distance will not much exceed 34 millions of miles.

But although this is a minimum distance for Mars, it must not be forgotten that it is a really vast distance, absolutely speaking; it is something like 150 times greater than the distance of the moon. With no telescopic power at our command could we possibly see anything on the moon of the size of the largest buildings or other works of human intelligence; so that we seem forever barred from detecting anything of the sort on Mars.

Nevertheless, the closest scrutiny of the ruddy planet by observers of great enthusiasm and intelligence, coupled with imagination and persistence, have built up a system of canals on Mars, covering the surface of the planet like spider webs over a printed page, crossing each other at intersecting spots known as "lakes," and embodying a wealth of detail which challenges criticism and explanation.

To see the canals at all requires a favorable presentation of Mars, a steady atmosphere and a perfect telescope, with a trained eye behind it. Not even then are they sure to be visible. The training of the eye has no doubt much to do with it. So photography has been called in, and very excellent pictures of Mars have already been taken, some nearly half as large as a dime, showing plainly the lights and shades of the grander divisions of the Martian surface, but only in a few instances revealing the actual canals more unmistakably than they are seen at the eyepiece.

The appearance and degree of visibility of the canals are variable: possibly clouds temporarily obscure them. But there is a certain capriciousness about their visibility that is little understood. In consequence of the changing physical aspects, as to season, on Mars and his orbital position with reference to the earth, some of the canals remain for a long time invisible, adding to the intricacy of the puzzle.

For the most part the canals are straight in their course and do not swerve much from a great circle on the planet. But their lengths are very different, some as short as 250 miles, some as long as 4,000 miles; and they often join one another like spokes in the hub of a wheel, though at various angles. As depicted by Lowell and his corps of observers at Flagstaff, Arizona, the canal system is a truly marvelous network of fine darkish stripes. Their color is represented as a bluish green.

Each marking maintains its own breadth throughout its entire length, but the breadth of all the canals is by no means the same: the narrowest are perhaps fifteen to twenty miles wide, and the broadest probably ten times that. At least that must be the breadth of the Nilosyrtis, which is generally regarded as the most conspicuous of all the canals. The Lowell Observatory has outstripped all others in the number of canals seen and charted, now about 500.

What may be the true significance of this remarkable system of markings it is impossible to conclude at present. Schiaparelli from his long and critical study of them, their changes of width and color, was led to think that they may be a veritable hydrographic system for distributing the liquid from the melting polar snows. In this case it would be difficult to escape the conviction that the canals have, at least in part, been designed and executed with a definite end in view.

Lowell went even farther and built upon their behavior an elaborate theory of life on the planet, with intelligent beings constructing and opening new canals on Mars at the present epoch. Pickering propounded the theory that the canals are not water-bearing channels at all, but that they are due to vegetation, starting in the spring when first seen and vitalized by the progress of the season poleward, the intensity of color of the vegetation coinciding with the progress of the season as we observe it.

Extensive irrigation schemes for conducting agricultural operations on a large scale seem a very plausible explanation of the canals, especially if we regard Mars as a world farther advanced in its life history than our own. Erosion may have worn the continents down to their minimum elevation, rendering artificial waterways not difficult to build; while with the vanishing Martian atmosphere and absence of rains, the necessity of water for the support of animal and vegetal life could only be met by conducting it in artificial channels from one region of the planet to another.

Interesting as this speculative interpretation is, however, we cannot pass by the fact that many competent astronomers with excellent instruments finely located have been unable to see the canals, and therefore think the astronomers who do see them are deceived in some way. Also many other astronomers, perhaps on insufficient grounds, deny their existence *in toto*.

Many patient years of labor would be required to consult all the literature of investigation of the planet Mars, but much of the detail has been critically embodied in maps at different epochs, by Kayser, Proctor, Green, and Dreyer. And Flammarion in two classic volumes on Mars has presented all the observations from the earliest time, together with his own interpretation of them. Areography is a term sometimes applied to a description of the surface of Mars, and it is scarcely an exaggeration to say that areography is now better known than the geography of immense tracts of the earth.

For some reason well recognized, though not at all well understood, Mars although the nearest of all the planets, Venus alone excepted, is an object by no means easy to observe with the telescope. Possibly its unusual tint has something to do with this. With an ordinary opera glass examine the moon very closely, and try to settle precise markings, colors, and the nature of objects on her surface; Mars under the best conditions, scrutinized with our largest and best telescope, presents a problem of about the same order of difficulty. There are delicate and changing local colors that add much uncertainty. Nevertheless, the planet's leading features are well made out, and their stability since the time of the earliest observers leaves no room to doubt their reality as parts of a permanent planetary crust.

The border of the Martian disk is brighter than the interior, but this brightness is far from uniform. Variations in the color of the markings often depend on the planet's turning round on its axis, and the relation of the surface to our angle of vision. If we keep in mind these obstacles to perfect vision in our own day, it is easy to see why the early users of very imperfect telescopes failed to see very much, and were misled by much that they thought they saw. Then, too, they had to contend, as we do, with unsteadiness of atmosphere, which is least troublesome near the zenith.

As their telescopes were all located in the northern hemisphere, the northern hemisphere of Mars is the one best circumstanced for their investigation; because at the remote oppositions of Mars, which always happen in our northern winter with the planet in high north declination, it is always the north pole of Mars which is presented to our view. Whereas the close oppositions of the planet always come in our northern midsummer, with Mars in south declination and therefore passing through the zenith of places in corresponding south latitude.

With Mars near opposition, high up from the horizon, a fairly steady atmosphere, and a magnifying power of at least 200 diameters, even the most casual observer could not fail to notice the striking difference in brightness of the two hemispheres: the northern chiefly bright and the southern markedly dark. Formerly this was thought to indicate that the southern hemisphere of Mars was chiefly water and the northern land, much as is the case on the earth: with this difference, however, that water and land on the earth are proportioned about as eleven to four.

But Mars in its general topography presents no analogy with the present relation of land and water on the earth. There seems no reason to doubt that the northern regions with their prevailing orange tint, in some places a dark red and in others fading to yellow and white, are really continental in character. Other vast regions of the Martian surface are possibly marshy, the varying depth of water causing the diversity of color. If we could ever catch a reflection of sunlight from any part of the surface of Mars, we might conclude that deep water exists on the planet; but the farther research progresses, the more complete becomes the evidence that permanent water areas on Mars, if they exist at all, are extremely limited.

Since 1877 Mars has been known to possess two satellites, which were discovered in August of that year by Hall at Washington. Moons of this planet had long been suspected to exist and on one or two previous occasions critically looked for, though without success. In the writings of Dean Swift there is a fanciful allusion to the two moons of Mars; and if astronomers had chanced to give serious attention to this, Phobos and Deimos, as Hall named them, might have been discovered long before.

They are very small bodies, not only faint in the telescope, but actually of only ten or twenty miles diameter; and from the strange relation that Phobos, the inner moon, moves round Mars three times while the planet itself is turning round only once on its axis, some astronomers incline to the hypothesis that this moon at least was never part of Mars itself, but that it was originally an inner or very eccentric member of the asteroid group, which ventured within the sphere of gravitation of Mars, was captured by that planet, and has ever since been tributary to it as a secondary body or satellite.

CHAPTER XXXIV
LIFE IN OTHER WORLDS

Popular interest in astronomy is exceedingly wide, but it is very largely confined to the idea of resemblances and differences between our earth and the bodies of the sky. The question most frequently asked the astronomer is, "Have any of the stars got people on them?" Or more specifically, "Is Mars inhabited?" The average questioner will not readily be turned off with yes or no for an answer. He may or may not know that it is quite impossible for astronomers to ascertain anything definite in this matter, most interesting as it is. What he wants to find out is the view of the individual astronomer on this absorbing and ever recurring inquiry.

We ought first to understand what is meant by the manifestation here on the earth called life, and agree concerning the conditions that render it possible. Apparently they are very simple. We may or may not agree that a counterpart of life, or life of a wholly different type from ours, may exist on other planets under conditions wholly diverse from those recognized as essential to its existence here. The problem of the origin of life is, in the present state of knowledge, highly speculative and hardly within the domain of science. Here on earth, life is intimately associated with certain chemical compounds, in which carbon is the common element without which life would not exist. Also hydrogen, oxygen, and nitrogen are present, with iron, sulphur, phosphorus, magnesium and a few less important elements besides. But carbon is the only substance absolutely essential. Protoplasm cannot be built without it, and protoplasm makes up the most of the living cell. Closely related to carbon is silica also, as a substitution in certain organic compounds. Protoplasm is able to stand very low temperatures, but its properties as a living cell cease when the temperature reaches 150 Fahrenheit.

Animal life as it exists on the earth to-day appears to have been here many million years. The palæontologists agree that all life originated in the waters of the earth. It has passed through evolutionary stages from the lowest to the highest. Throughout this vast period the astronomer is able to say that the conditions of the earth which appear to be essential to the maintenance of life have been pretty constantly what they are to-day. The higher the type of life, the narrower the range of conditions under which it thrives. Man can exist at the frigid poles even if the temperature is 75 degrees below Fahrenheit zero; and in the deserts and the tropics, he swelters under temperatures of 115 degrees, but he still lives. At these extremes, however, he can scarcely be said to thrive.

We have, then, a relatively narrow range of temperatures which seems to be essential to his comfortable existence and development: we may call it 150 degrees in extent. Had not the surface temperature of the earth been maintained within this range for indefinite ages, in the regions where the human race has developed, quite certainly man would not be here. How this equability of temperature has been maintained does not now matter. Clearly the earth must have existed through indefinite ages in the process of cooling down from temperatures of at least 6,000 degrees.

During this stage the temperature of the surface was earth-controlled. Then this period merged very gradually into the stage where life became possible, and the temperature of the surface became, as it now is, sun-controlled. How many years are embraced in this span of periods, or ages, we have no means of knowing. But of the sequence of periods and the secular diminution of temperature, we may be certain.

Then there is the equally important consideration of water necessary for the origination, support, and development of life. We cannot conceive of life existing without it. On the earth water is superabundant, and has been for indefinite ages in the past. There is little evidence that the oceans are drying up; although the commonly accepted view is that the waters of the earth will very gradually disappear. Water can exist in the fluid state, which is essential to life, at all temperatures between 32 degrees and 680 degrees F.

Air to breathe is essential to life also. The atmosphere which envelops the earth is at least 100 miles in depth, and its own weight compresses it to a tension of nearly 15 pounds to the square inch at sea level. This atmosphere and its physical properties have had everything to do with the development of animal life on the planet. Without it and its remarkable property of selective absorption, which imprisons and diffuses the solar heat, it is inconceivable that the necessary equability of surface temperature could be maintained. This appears to be quite independent of the chemical constituents of the atmosphere, and is perhaps the most important single consideration affecting the existence of life on a planet. If the surface of a planet is partly covered with water, it will possess also an atmosphere containing aqueous vapor.

Heat, water, and air: these three essentials determine whether there is life on a planet or not. Of course there must be nutrition suitable to the organism; mineral for the vegetal, and vegetal for the animal. But the narrow range of variation appears to be the striking thing: relatively but a few degrees of temperature, and a narrow margin of atmospheric pressure. If this pressure is doubled or trebled, as in submarine caissons, life becomes insupportable. If, on the other hand, it is reduced even one-third, as on mountains even 13,000 feet high, the human mechanism fails to function, partly from lack of oxygen necessary in vitalizing the blood, but mainly because of simple reduction of mechanical pressure.

If, then, we conceive of life in other worlds and it is agreed that life there must manifest itself much as it does here, our answer to the question of habitability of the planets must follow upon an investigation of what we know, or can reasonably surmise, about the surface temperatures of these bodies, whether they have water, and what are the probable physical characteristics of their atmospheres.

We may inquire about each planet, then, concerning each of these details.

The case of Mercury is not difficult. At an average distance of only 36 million miles from the sun, and with a large eccentricity of orbit which brings it a fifth part nearer, conditions of temperature alone must be such as to forbid the existence of life. The solar heat received is seven times greater than at the earth, and this is perhaps sufficient reason for a minimum of atmosphere, as indicated by observation. If no air, then quite certainly no water, as evaporation would supply a slight atmosphere. But according to the kinetic theory of gases, the mass of Mercury, only a very small fraction of that of the sun, is inadequate to retain an atmospheric envelope. If, however, the planet's day and year are equal, so that it turns a constant face to the sun, surface conditions would be greatly complicated, so that we cannot regard the planet as absolutely uninhabitable on the hemisphere that is always turned away from the sun.

Venus at 67 millions of miles from the sun presents conditions that are quite different. She receives double the solar heat that we do, but possessing an atmosphere perhaps threefold denser than ours, as reliably indicated by observations of transits of Venus, the intensity of the heat and its diffusion may be greatly modified. What the selective absorption of the atmosphere of Venus may be, we do not know. Nor is the rotation time of the planet definitely ascertained: if equal to her year, as many observations show and as indicated by the theory of tidal evolution, there may well be certain regions on the hemisphere perpetually turned away from the sun where temperature conditions are identical with those on the tropical earth, and where every condition for the origin and development of life is more fully met than anywhere else in the solar system. Whether Venus has water distributed as on the earth we do not know, as her surface is never seen, owing to dense clouds under which she is always enshrouded. Her cloudy condition possibly indicates an overplus of water.

Is the moon inhabited? Quite certainly not: no appreciable air, no water, and a surface temperature unmodified by atmosphere—rising perhaps to 100 degrees F. during the day, which is a fortnight in length, and falling at night to 300 degrees below zero, if not lower.

Is Mars inhabited? The probable surface temperature is much lower than the earth's, because Mars receives only half as much solar heat as we do; and more important still, the atmosphere of Mars is neither so dense nor so extensive as our own. Seasons on Mars are established, much the same as here, except that they are nearly twice as long as ours; and alternate shrinking and enlarging of the polar caps keeps even pace with the seasons, thereby indicating a certainty of atmosphere whose equatorial and polar circulation transports the moisture poleward to form the snow and ice of which the polar caps no doubt consist.

There is a variety of evidence pointing to an atmosphere on Mars of one-third to one-half the density of our own: an atmosphere in which free hydrogen could not exist, although other gases might. The spectroscopic evidence of water vapor in the Martian atmosphere is not very strong. It is very doubtful whether water exists on Mars in large bodies: quite certainly not as oceans, though the evidence of many small "lakes" is pretty well made out. With very little water, a thin atmosphere and a zero temperature, is Mars likely to be inhabited at the present time? The chances are rather against it. If, however, the past development of the planet has progressed in the way usually considered as probable, we may be practically certain that Mars has been inhabited in the past, when water was more abundant, and the atmosphere more dense so as to retain and diffuse the solar heat.

Biologists tell me that they hardly know enough regarding the extreme adaptability of organisms to environment to enable them to say whether life on such a planet as Mars would or would not keep on functioning with secular changes of moisture and temperature. The survival of a race might be insured against extremely low temperatures by dwelling in sub-Martian caves, and sufficient water might be preserved by conceivable engineering and mechanical schemes; but the secular reduction of the quantity and pressure of atmosphere—it is not easy to see how a race even more advanced than ourselves could maintain itself alive under serious lack of an element so vital to existence. Both Wallace, the great biologist, and Arrhenius, the eminent chemist (but biologist, astronomer, and physicist as well), both reject the habitation theory of Mars, regarding the so-called canals as quite like the luminous streaks on the moon; that is, cracks in the volcanic crust caused by internal strains due to the heated interior. Wallace, indeed, argues that the planet is absolutely uninhabitable.

The asteroids, or minor planets? We may dismiss them with the simple consideration that their individual masses are so insignificant and their gravity so slight that no atmosphere can possibly surround them. Their temperatures must be exceedingly low, and water, if present at all, can only exist in the form of ice.

Jupiter, the giant planet, presents the opposite extreme. His mass is nearly a thousandth part of the sun's, and is sufficient to retain a very high temperature, probably approximating to the condition we call red-hot. This precludes the possibility of life at the outset, although the indications of a very dense atmosphere many thousand miles in depth are unmistakable.

Of Saturn, one thirty-five hundredth the mass of the sun, practically the same may be said. Proctor thought it quite likely that Saturn might be habitable for living creatures of some sort, but he regarded the planet as on many accounts unsuitable as a habitation for beings constituted like ourselves. Mere consideration of surface temperature precludes the possibility of life in the present stage of Saturn's development; but the consensus of opinion is to the effect that life may make its appearance on these great planets at some inconceivably remote epoch in the future when the surface temperature is sufficiently reduced for life processes to begin. Discoveries of algæ flourishing in hot springs approaching 200 degrees Fahrenheit make it possible that these beginnings may take place earlier and at much higher temperatures than have hitherto been thought possible.

A century ago, when the ring of Saturn was believed to be a continuous plane, this was a favorite corner of the solar system for speculation as to habitability; but now that we know the true constitution of the rings, no one would for a moment consider any such possibility. Conditions may, however, be quite different with Saturn's huge satellite Titan, the giant moon of the solar system. Its diameter makes it approximately the size of the planet Mars; and although it is much farther removed from the sun, its relative nearness to the highly heated globe of Saturn may provide that equability of temperature which is essential to life processes.

Also the three inner Galilean moons of Jupiter, especially III which is about the size of Titan, are excellently placed for life possibilities, as far as probable temperature is concerned, but we have of course no basis for surmising what their conditions may be as to air and water, except that their small mass would indicate a probable deficiency of those elements.

Uranus and Neptune are planets so remote, and their apparent disks are so small, that very little is known about their physical condition. They are each about one-third the diameter of Jupiter, and the spectrum of Uranus shows broad diffused bands, indicating strong absorption by a dense atmosphere very different from that of the earth. Indications are that Neptune has a similar atmosphere.

It is possible that the denser atmospheres of these remote planets may be so conditioned as to selective absorption that the relatively slender supply of solar heat may be conserved, and thus insure a relatively high surface temperature when the sun comes into control. If our theories of origin of the planets are to be trusted, we may rather suppose that Uranus and Neptune are still in a highly heated condition; that life has not yet made its appearance on them, but that it will begin its development ages before Saturn and Jupiter have cooled to the requisite temperature.

Comets? In his *Lettres Cosmologiques* (1765) Lambert considers the question of habitability of the comets, naturally enough in his day, because he thought them solid bodies surrounded by atmosphere, and related to the planets. The extremes of temperature at perihelia and aphelia to which comets are subjected did not bother him particularly.

After calculating that the comet of 1680, "being 160 times nearer to the sun than we are ourselves, must have been subjected to a degree of heat 25,600 times as great as we are," Lambert goes on to say: "Whether this comet was of a more compact substance than our globe, or was protected in some other way, it made its perihelion passage in safety, and we may suppose all its inhabitants also passed safely. No doubt they would have to be of a more vigorous temperament and of a constitution very different from our own. But why should all living beings necessarily be constituted like ourselves? Is it not infinitely more probable that amongst the different globes of the universe a variety of organizations exist, adapted to the wants of the people who inhabit them, and fitting them for the places in which they dwell, and the temperatures to which they will be subjected? Is man the only inhabitant of the earth itself? And if we had never seen either bird or fish, should we not believe that the air and water were uninhabitable? Are we sure that fire has not its invisible inhabitants, whose bodies, made of asbestos, are impenetrable to flame? Let us admit that the nature of the beings who inhabit comets is unknown to us; but let us not deny their existence, and still less the possibility of it."

Little enough is really known about the physical nature of comets even now, but what we do know indicates incessant transformation and instability of conditions that would render life of any type exceedingly difficult of maintenance.

A word about Sir William Herschel's theory of the sun and its habitability. He thought the core of the sun a dark, solid body, quite cold, and surrounded by a double layer, the inner one of which he conceived to act as a sort of fire screen to shield the sun proper against the intense heat of the outer layer, or photosphere by which we see it. Viewed in this light, the sun, he says, "appears to be nothing else than a very eminent, large and lucid planet, evidently the first, or, in strictness of speaking, the only primary one of our system.... It is most probably also inhabited, like the rest of the planets, by beings whose organs are adapted to the peculiar circumstances of that vast globe." But physics and biology were undeveloped sciences in Herschel's days.

Herschel knew, however, that the stars are all suns, so that he must have conceived that they are inhabited also, quite independently of the question whether they possess retinues of planets, after the manner of our solar system.

This again is a question to which the astronomer of the present day can give no certain answer. So immensely distant are even the nearest of these multitudinous bodies that no telescope can ever be built large enough or powerful enough to reveal a dark planet as large as Jupiter, alongside even the nearest fixed star. Whatever may be the process of stellar evolution, there doubtless is an era of many hundreds of millions of years in the life of a star when it is passing through a planet-maintaining stage. This would likely depend upon spectral type, or to be indicated by it; and as about half of the stars are of the solar type, it would be a reasonable inference that at least half of the stars may have planets tributary to them.

In such a case, the chances must be overwhelmingly in favor of vast numbers of the planets of other stellar systems being favorably circumstanced as to heat and moisture for the maintenance of life at the present time. That is, they are habitable, and if habitable, then thousands of them are no doubt inhabited now. But astronomers know absolutely nothing about this question, nor are they able to conceive at present any way that may lead them to any definite knowledge of it. There is, indeed, one piece of quasi-evidence which might reasonably be interpreted as implying that it is more likely that the stars are not attended by families of planets than that they are.

CHAPTER XXXV

THE LITTLE PLANETS

Along toward the end of the eighteenth century and the beginning of the nineteenth, astronomers were leading a quiet unexcited life. Sir William Herschel had been knighted by King George for his discovery of the outer planet Uranus, and practically everything seemed to be known and discovered in the solar system with a single exception. Between Mars and Jupiter there existed an obvious gap in the planetary brotherhood.

Could it be possible that some time in the remote cosmic past a planet had actually existed there, and that some celestial cataclysm had blown it to fragments? If so, would they still be traveling round the sun as individual small planets? And might it not be possible to discover some of them among the faint stars that make up the belt of the zodiac in which all the other planets travel?

So interesting was this question that the first international association of astronomers banded themselves together to carry on a systematic search round the entire zodiacal heavens in the faint hope of detecting possible fragments of the original planet of mere hypothesis.

The astronomers of that day placed much reliance on what is known as Bode's law—not a law at all, but a mere arithmetical succession of numbers which represented very well the relative distances of all the planets from the sun. And the distance of the newly found Uranus fitted in so well with this law that the utter absence of a planet in the gap between Mars and Jupiter became very strongly marked.

Quite by accident a discovery of one of the guessed-at small planetary bodies was made, on January 1, 1801, in Palermo, Sicily, by Piazzi, who was regularly occupied in making an extensive catalogue of the stars. His observations soon showed that the new object he had seen could not be a fixed star, because it moved from night to night among the stars. He concluded that it was a planet, and named it Ceres (1), for the tutelary goddess of Sicily.

Other astronomers kept up the search, and another companion planet, Pallas (2) was found in the following year. Juno (3) was found in 1804, and Vesta (4), the largest and brightest of all the minor planets, in 1807. Vesta is sometimes bright enough when nearest the earth to be seen with the naked eye; but it was the last of the brighter ones, and no more discoveries of the kind were made till the fifth was found in 1845. Since then discoveries have been made in great abundance, more and more with every year till the number of little planets at present known is very near 1,000.

The early asteroid hunters found the search rather tedious, and the labor increased as it became necessary to examine the increasing thousands of fainter and fainter stars that must be observed in order to detect the undiscovered planets, which naturally grow fainter and fainter as the chase is prolonged. First a chart of the ecliptic sky had to be prepared containing all the stars that the telescope employed in the search would show. Some of the most detailed charts of the sky in existence were prepared in connection with this work, particularly by the late Dr. Peters of Hamilton College. Once such charts are complete, they are compared with the sky, night after night when the moon is absent. Thousands upon thousands of tedious hours are spent in this comparison, with no result whatever except that chart and sky are found to correspond exactly.

But now and then the planet hunter is rewarded by finding a new object in the sky that does not appear on his chart. Almost certainly this is a small planet, and only a few night's observation will be necessary to enable the discoverer to find out approximately the orbit it is traveling in, and whether it is out-and-out a new planet or only one that had been previously recognized, and then lost track of.

Nearly all the minor planets so far found have had names assigned to them principally legendary and mythological, and a nearly complete catalogue of them, containing the elements of their orbits (that is, all the mathematical data that tell us about their distance from the sun and the circumstances of their motion around him) is published each year in the "Annuaire du Bureau des Longitudes" at Paris. But these little planets require a great deal of care and attention, for some astronomers must accurately observe them every few years, and other astronomers must conduct intricate mathematical computations based on these observations; otherwise they get lost and have to be discovered all over again. Professor Watson, of the University of Michigan and later of the University of Wisconsin, endowed the 22 asteroids of his own discovery, leaving to the National Academy of Sciences a fund for prosecuting this work perpetually, and Leuschner is now ably conducting it.

While the number of the asteroids is gratifyingly large, their individual size is so small and their total mass so slight that, even if there are a hundred thousand of them (as is wholly possible), they would not be comparable in magnitude with any one of the great planets. Vesta, the largest, is perhaps 400 miles in diameter, and if composed of substances similar to those which make up the earth, its mass may be perhaps one twenty-thousandth of the earth's mass. If we calculate the surface gravity on such a body, we find it about one-thirtieth of what it is here; so that a rifle ball, if fired on Vesta with a muzzle velocity of only 2,000 feet a second, might overmaster the gravity of the little planet entirely and be projected in space never to return.

If, as is likely, some of the smallest asteroids are not more than ten miles in diameter, their gravity must be so feeble a force that it might be overcome by a stone thrown from the hand. There is no reliable evidence that any of the asteroids are surrounded by atmospheric gases of any sort. Probably they are for the most part spherical in form, although there is very reliable evidence that a few of the asteroids, being variable in the amount of sunlight that they reflect, are irregular in form, mere angular masses perhaps.

Jupiter, Largest of the Planets. The irregular belts change their mutual relation and shapes because they do not represent land, but are part of the atmosphere.

(Photo, Yerkes Observatory.)

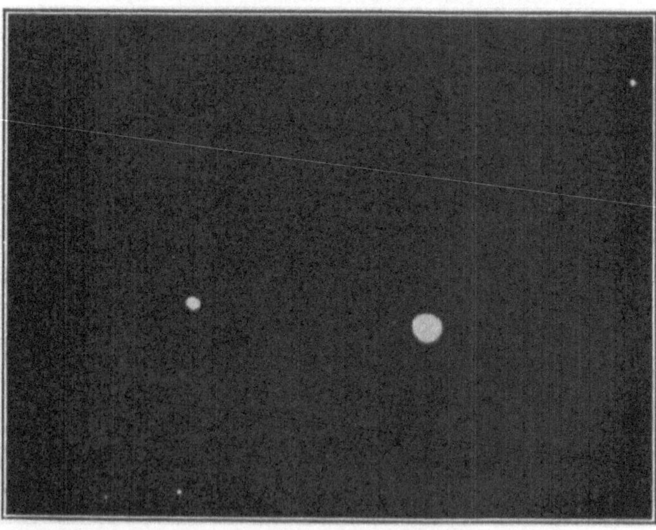

The Planet Neptune and its Satellite. The photograph required an exposure of the plate for one hour.

(Photo, Yerkes Observatory.)

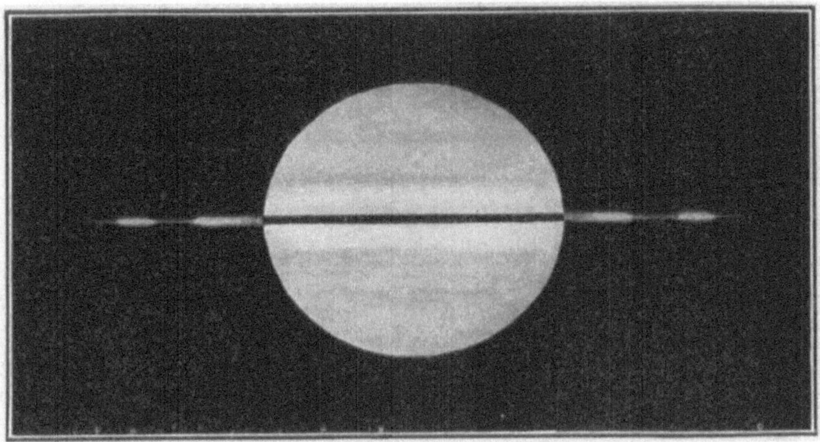

Saturn, as Seen Through the 40-inch Refractor, at the time when only the edge of the rings is visible, showing condensations.

(Photo, Yerkes Observatory.)

Saturn, Photographed Through the 40-inch Refractor. The rings appear opened to the fullest extent they can be seen from the earth. The picture was made July 7, 1898.

(Photo, Yerkes Observatory.)

The network of orbits of the asteroids is inconceivable complicated. Nevertheless, there is a wide variation in their average distance from the sun, and their periods of traveling round him vary in a similar manner, the shortest being only about three years. While the longest is nearly nine years in duration, the average of all their periods is a little over four years. The gap in the zone of asteroids, at a distance from the sun equal to about five-eighths that of Jupiter, is due to the excessive disturbing action of Jupiter, whose periodic time is just twice as long as that of a theoretical planet at this distance.

The average inclination of their orbits to the plane of the ecliptic is not far from 8 degrees. But the orbit of Pallas, for example, is inclined 35 degrees, and the eccentricities of the asteroid orbits are equally erratic and excessive. Both eccentricity and inclination of orbit at times suggest a possible relation to cometary orbits, but nothing has ever been definitely made out connecting asteroids and comets in a related origin.

No comprehensive theory of the origin of the asteroid group has yet been propounded that has met with universal acceptance. According to the nebular hypothesis the original gaseous material, which should have been so concentrated as to form a planet of ordinary type, has in the case of the asteroids collected into a multitude of small masses instead of simply one. That there is a sound physical reason for this can hardly be denied. According to the Laplacian hypothesis, the nearness of the huge planetary mass of Jupiter just beyond their orbits produced violent perturbations which caused the original ring of gaseous material to collect into fragmentary masses instead of one considerable planet. The theory of a century ago that an original great planet was shattered by internal explosive forces is no longer regarded as tenable.

To astronomers engaged upon investigation of distances in the solar system, the asteroid group has proved very useful. The late Sir David Gill employed a number of them in a geometrical research for finding the sun's distance, and more recently the discovery of Eros (433) has made it possible to apply a similar method for a like purpose when it approaches nearest to the earth in 1924 and 1931. Then the distance of Eros will be less than half that of Mars or even Venus at their nearest.

When the total number of asteroids discovered has reached 1,000, with accurate determination of all their orbits, we shall have sufficient material for a statistical investigation of the group which ought to elucidate the question of its origin, and bear on other problems of the cosmogony yet unsolved. Present methods of discovery of the asteroids by photography replace entirely the old method by visual observation alone, with the result that discoveries are made with relatively great ease and rapidity.

CHAPTER XXXVI

THE GIANT PLANET

I can never forget as a young boy my first glimpse of the planet Jupiter and his moons; it was through a bit of a telescope that I had put together with my own hands; a tube of pasteboard, and a pair of old spectacle lenses that chanced to be lying about the house.

In the field of view I saw five objects; four of them looking quite alike, and as if they were stars merely (they were Jupiter's moons), while the fifth was vastly larger and brighter. It was circular in shape, and I thought I could see a faint darkish line across the middle of it.

This experience encouraged me immensely, and I availed myself eagerly of the first chance to see Jupiter through a bigger and better glass. Then I saw at once that I had observed nothing wrongly, but that I had seen only the merest fraction of what there was to see.

In the first place, the planet's disk was not perfectly circular, but slightly oval. Inquiring into the cause of this, we must remember that Jupiter is actually not a flat disk but a huge ball or globe, more than ten times the diameter of the earth, which turns swiftly round on its axis once every ten hours as against the earth's turning round in twenty-four hours. Then it is easy to see how the centrifugal force bulges outward the equatorial regions of Jupiter, so that the polar regions are correspondingly drawn inward, thereby making the polar diameter shorter than the equatorial one, which is in line with the moons or satellites. The difference between the two diameters is very marked, as much as one part in fifteen. All the planets are slightly flattened in this way, but Jupiter is the most so of all except Saturn.

The little darkish line across the planet's middle region or equator was found to be replaced by several such lines or irregular belts and spots, often seen highly colored, especially with reflecting telescopes; and they are perpetually changing their mutual relation and shapes, because they are not solid territory or land on Jupiter, but merely the outer shapes of atmospheric strata, blown and torn and twisted by atmospheric circulation on this planet, quite the same as clouds in the atmosphere on the earth are.

Besides this the axial turning of Jupiter brings an entirely different part of the planet into view every two or three hours; so that in making a map or chart of the planet, an arbitrary meridian must be selected. Even then the process is not an easy one, and it is found that spots on Jupiter's equator turn round in 9 hours 50 minutes, while other regions take a few minutes longer, the nearer the poles are approached. The Great Red Spot, about 30,000 miles long and a quarter as much in breadth has been visible for about half a century. Bolton, an English observer, has made interesting studies of it very recently.

The four moons, or satellites, which a small telescope reveals, are exceedingly interesting on many accounts. They were the first heavenly bodies seen by the aid of the telescope, Galileo having discovered them in 1610. They travel round Jupiter much the same as the moon does round the earth, but faster, the innermost moon about four times per week, the second moon about twice a week, the third or largest moon (larger than the planet Mercury) once a week, and the outermost in about sixteen days. The innermost is about 260,000 miles from Jupiter, and the outermost more than a million miles. From their nearness to the huge and excessively hot globe of Jupiter, some astronomers, Proctor especially, have inclined to the view that these little bodies may be inhabited.

Jupiter has other moons; a very small one, close to the planet, which goes round in less than twelve hours, discovered by Barnard in 1892. Four others are known, very small and faint and remote from the planet, which travel slowly round it in orbits of great magnitude. The ninth, or outermost, is at a distance of fifteen and one-half million miles from Jupiter, and requires nearly three years in going round the planet. It was discovered by Nicholson at the Lick Observatory in 1914. The eighth was discovered by Melotte at Greenwich in 1908, and is peculiar in the great angle of 28 degrees, at which its orbit is inclined to the equator of Jupiter. The sixth and seventh satellites revolve round Jupiter inside the eighth satellite, but outside the orbit of IV; and they were discovered by photography at the Lick Observatory in 1905 by Perrine, now director of the Argentine National Observatory at Cordoba.

The ever-changing positions of the Medicean moons, as Galileo called the four satellites that he discovered—their passing into the shadow in eclipse, their transit in front of the disk, and their occultation behind it—form a succession of phenomena which the telescopist always views with delight. The times when all these events take place are predicted in the "Nautical Almanac," many thousand of them each year, and the predictions cover two or three years in advance.

Jupiter, as the naked eye sees him high up in the midnight sky, is the brightest of all the planets except Venus; indeed, he is five times brighter than Sirius, the brightest of all the fixed stars. His stately motion among the stars will usually be visible by close observation from day to day, and his distance from the earth, at times when he is best seen, is usually about 400 million miles. Jupiter travels all the way round the sun in twelve years; his motion in orbit is about eight miles a second.

The eclipses of Jupiter's moons, caused by passing into the shadow of the planet, would take place at almost perfectly regular intervals, if our distance from Jupiter were invariable. But it was early found out that while the earth is approaching Jupiter the eclipses take place earlier and earlier, but later and later when the earth is moving away. The acceleration of the earliest eclipse added to the retardation of the latest makes 1,000 seconds, which is the time that light takes in crossing a diameter of the earth's orbit round the sun. Now the velocity of light is well known to be 186,300 miles per second, so we calculate at once and very simply that the sun's distance from the earth, which is half the diameter of the orbit, equals 500 times 186,300, or 93,000,000 miles.

CHAPTER XXXVII
THE RINGED PLANET

Saturn is the most remote of all the planets that the ancient peoples knew anything about. These anciently known planets are sometimes called the lucid or naked-eye planets—five in number: Mercury, Venus, Mars, Jupiter, and Saturn. Saturn shines as a first-magnitude star, with a steady straw-colored light, and is at a distance of about 800 million miles from the earth when best seen. Saturn travels completely round the sun in a little short of thirty years, and the telescope, when turned to Saturn, reveals a unique and astonishing object; a vast globe somewhat similar to Jupiter, but surrounded by a system of rings wholly unlike anything else in the universe, as far as at present known; the whole encircled by a family of ten moons or satellites. The Saturnian system, therefore, is regarded by many as the most wonderful and most interesting of all the objects that the telescope reveals.

At first the flattening of the disk of Saturn is not easily made out, but every fifteen years (as 1921 and 1936) the earth comes into a position where we look directly at the thin edge of the rings, causing them to completely disappear. Then the remarkable flattening of the poles of Saturn is strikingly visible, amounting to as much as one-tenth of the entire diameter. The atmospheric belt system is also best seen at these times.

But the rings of Saturn are easily the most fascinating features of the system. They can never be seen as if we were directly above or beneath the planet so they never appear circular, as they really are in space, but always oval or elliptical in shape. The minor axis or greatest breadth is about one-half the major axis or length. The latter is the outer ring's actual diameter, and it amounts to 170,000 miles, or two and one-half times the diameter of Saturn's globe.

There are in fact no less than four rings; an outer ring, sometimes seen to be divided near its middle; an inner, broader and brighter ring; and an innermost dusky, or crape ring, as it is often called. This comes within about 10,000 miles of the planet itself. After the form and size of the rings were well made out, their thickness, or rather lack of thickness, was a great puzzle.

If a model about a foot in diameter were cut out of tissue paper, the relative proportion of size and thickness would be about right. In space the thickness is very nearly 100 miles, so that, when we look at the ring system edge-on, it becomes all but invisible except in very large telescopes. Clearly a ring so thin cannot be a continuous solid object and recent observations have proved beyond a doubt that Saturn's rings are made up of millions of separate particles moving round the planet, each as if it were an individual satellite.

Ever since 1857 the true theory of the constitution of the Saturnian ring has been recognized on theoretic grounds, because Clerke-Maxwell founded the dynamical demonstration that the rings could be neither fluid nor solid, so that they must be made up of a vast multitude of particles traveling round the planet independently. But the physical demonstration that absolutely verified this conclusion did not come until 1895, when, as we have said in a preceding chapter, Keeler, by radial velocity measures on different regions of the ring by means of the spectroscope, proved that the inner parts of the ring travel more swiftly round the planet than the outer regions do. And he further showed that the rates of revolution in different parts of the ring exactly correspond to the periods of revolution which satellites of Saturn would have, if at the same distance from the center of the planet. The innermost particles of the dusky ring, for example, travel round Saturn in about five hours, while the outermost particles of the outer bright ring take 137 hours to make their revolution. For many years it was thought that the Saturnian ring system was a new satellite in process of formation, but this view is no longer entertained; and the system is regarded as a permanent feature of the planet, although astronomers are not in entire agreement as to the evolutionary process by which it came into existence—whether by some cosmic cataclysm, or by gradual development throughout indefinite aeons, as the rest of the solar system is thought to have come to its present state of existence. Possibly the planetesimal hypothesis of Chamberlin and Moulton affords the true explanation, as the result of a rupture due to excessive tidal strain.

CHAPTER XXXVIII

THE FARTHEST PLANETS

On the 13th of March, 1781, between 10 and 11 P. M., as Sir William Herschel was sweeping the constellation Gemini with one of his great reflecting telescopes, one star among all that passed through the field of view attracted his attention. Removing the eyepiece and applying another with a higher magnifying power, he found that, unlike all the other stars, this one had a small disk and was not a mere point of light, as all the fixed stars seem to be.

A few nights' observation showed that the stranger was moving among the stars, so he thought it must be a comet; but a week's observation following showed that he had discovered a new member of the planetary system, far out beyond Saturn, which from time immemorial had been assumed to be the outermost planet of all. This, then, was the first real discovery of a planet, as the finding of the satellites of Jupiter had been the first of all astronomical discoveries. Herschel's discovery occasioned great excitement, and he named the new planet Georgium Sidus or the Georgian, after his King. The King created him a knight and gave him a pension, besides providing the means for building a huge telescope, 40 feet long, with which he subsequently made many other astronomical discoveries. The planet that Herschel discovered is now called Uranus.

Uranus is an object not wholly impossible to see with the naked eye, if the sky background is clear and black, and one knows exactly where to look for it. Its brightness is about that of a sixth magnitude star or a little fainter. Its average distance from the sun is about 1,800 million miles and it takes eighty-four years to complete its journey round the sun, traveling only a little more than four miles a second. When we examine Uranus closely with a large telescope, we find a small disk slightly greenish in tint, very slightly flattened, and at times faint bands or belts are apparently seen. Uranus is about 30,000 miles in diameter, and is probably surrounded by a dense atmosphere. Its rotation time is 10 h. 50 m.

Uranus is attended by four moons or satellites, named Ariel, Umbriel, Titania, and Oberon, the last being the most remote from the planet. This system of satellites has a remarkable peculiarity: the plane of the orbits in which they travel round Uranus is inclined about 80 degrees to the plane of the ecliptic, so that the satellites travel backward, or in a retrograde direction; or we might regard their motion as forward, or direct, if we considered the planes of their orbits inclined at 100 degrees.

For many years after the discovery of Uranus it was thought that all the great bodies of the solar system had surely been found. Least of all was any planet suspected beyond Uranus until the mathematical tables of the motion of Uranus, although built up and revised with the greatest care and thoroughness, began to show that some outside influence was disturbing it in accordance with Newton's law of gravitation. The attraction of a still more distant planet would account for the disturbance, and since no such planet was visible anywhere a mathematical search for it was begun.

NEPTUNE

Wholly independently of each other, two young astronomers, Adams of England and Le Verrier of France, undertook to solve the unique problem of finding out the position in the sky where a planet might be found that would exactly account for the irregular motion of Uranus. Both reached practically identical results. Adams was first in point of time, and his announcement led to the earliest observation, without recognition of the new planet (July 30, 1846), although it was Le Verrier's work that led directly to the new planet's being first seen and recognized as such (September 23, 1846). Figuring backward, it was found that the planet had been accidentally observed in Paris in 1795, but its planetary character had been overlooked.

Neptune is the name finally assigned to this historical planet. It is thirty times farther from the sun than the earth, or 2,800 million miles; its velocity in orbit is a little over three miles per second, and it consumes 164 years in going once completely round the sun. So faint is it that a telescope of large size is necessary to show it plainly. The brightness equals that of a star of the eighth magnitude, and with a telescope of sufficient magnifying power, the tiny disk can be seen and measured. The planet is about 30,000 miles in diameter, and is not known to possess more than one moon or satellite. If there are others, they are probably too faint to be seen by any telescope at present in existence.

CHAPTER XXXIX

THE TRANS-NEPTUNIAN PLANET

Investigation of the question of a possible trans-Neptunian planet was undertaken by the writer in 1877. As Neptune requires 164 years to travel completely round the sun, and the period during which it has been carefully observed embraces only half that interval, clearly its orbit cannot be regarded as very well known. Any possible deviations from the mathematical orbit could not therefore be traced to the action of a possible unknown planet outside. But the case was different with Uranus, which showed very slight disturbances, and these were assumed to be due to a possible planet exterior to both Uranus and Neptune. As a position for this body in the heavens was indicated by the writer's investigation, that region of the sky was searched by him with great care in 1877-1878 with the twenty-six-inch telescope at Washington; and photographs of the same region were afterward taken by others, though only with negative results.

In 1880, Forbes of Edinburgh published his investigation of the problem from an entirely independent angle. Families of comets have long been recognized whose aphelion distances correspond so nearly with the distances of the planets that these comet families are now recognized as having been created by the several planets, which have reduced the high original velocities possessed by the comets on first entering the solar system.

Their orbits have ever since been ellipses with their aphelia in groups corresponding to the distances of the planets concerned. Jupiter has a large group of such comets, also Saturn. Uranus and Neptune likewise have their families of comets, and Forbes found two groups with average distances far outside of Neptune; from which he drew the inference that there are two trans-Neptunian planets. The position he assigned to the inner one agreed fairly well with the writer's planet as indicated by unexplained deviations of Uranus.

The theoretical problem of a trans-Neptunian planet has since been taken up by Gaillot and Lau of Paris, the late Percival Lowell, and W. H. Pickering of Harvard. The photographic method of search will, it is expected, ultimately lead to its discovery. On account of the probable faintness of the planet, at least the twelfth or thirteenth magnitude, Metcalf's method of search is well adapted to this practical problem. When near its opposition the motion of Neptune retrograding among the stars amounts to five seconds of arc in an hour; while the trans-Neptunian planet would move but three seconds. By shifting the plate this amount hourly during exposure, the suspected object would readily be detected on the photographic plate as a minute and nearly circular disk, all the adjacent stars being represented by short trails.

Interest in a possible planet or planets outside the orbit of Neptune is likely to increase rather than diminish. To the ancients seven was the perfect number, there were seven heavenly bodies already known, so there could be no use whatever in looking for an eighth. The discovery of Uranus in 1781 proved the futility of such logic, and Neptune followed in 1846 with further demonstration, if need be. The cosmogony of the present day sets no outer limit to the solar system, and some astronomers advocate the existence of many trans-Neptunian planets.

CHAPTER XL

COMETS—THE HAIRY STARS

Comets—hairy stars, as the origin of the name would indicate—are the freaks of the heavens. Of great variety in shape, some with heads and some without, some with tails and some without, moving very slowly at one time and with exceedingly high velocity at another, in orbits at all possible angles of inclination to the general plane of the planetary paths round the sun, their antics and irregularities were the wonder and terror of the ancient world, and they are keenly dreaded by superstitious people even to the present day.

Down through the Middle Ages the advent of a comet was regarded as:

> Threatening the world with famine, plague and war;
> To princes, death; to kingdoms, many curses;
> To all estates, inevitable losses;
> To herdsmen, rot; to plowmen, hapless seasons;
> To sailors, storms; to cities, civil treasons.

Comets appeared to be marvelous objects, as well as sinister, chiefly because they bid apparent defiance to all law. Kepler had shown that the moon and the planets travel in regular paths—slightly elliptical to be sure, but nevertheless unvarying. None of the comets were known to follow regular paths till the time of Halley late in the seventeenth century, when, as we have before told, a fine comet made its appearance, and Halley calculated its orbit with much precision. Comparing this with the orbits of comets that had previously been seen, he found its path about the sun practically identical with that of at least two comets previously observed in 1531 and 1607.

So Halley ventured to think all these comets were one and the same body, and that it traveled round the sun in a long ellipse in a period of about seventy-five or seventy-six years. We have seen how his prediction of its return in 1758 was verified in every particular. On the comet's return in 1910, Crowell and Crommelin of Greenwich made a thorough mathematical investigation of the orbit, indicating that the year 1986 will witness its next return to the sun.

There is a class of astronomers known as comet-hunters, and they pass hours upon hours of clear, sparkling, moonless nights in search for comets. They are equipped with a peculiar sort of telescope called a comet-seeker, which has an object glass usually about four or five inches in diameter, and a relatively short length of focus, so that a larger field of view may be included. Regions near the poles of the heavens are perhaps the most fruitful fields for search, and thence toward the sun till its light renders the sky too bright for the finding of such a faint object as a new comet usually is at the time of discovery. Generally when first seen it resembles a small circular patch of faint luminous cloud.

When a suspect is found, the first thing to do is to observe its position accurately with relation to the surrounding stars. Then, if on the next occasion when it is seen the object has moved, the chances are that it is a comet; and a few days' observation will provide material from which the path of the comet in space can be calculated. By comparing this with the complete lists of comets, now about 700 in number, it is possible to tell whether the comet is a new one, or an old one returning. The total number of comets in the heavens must be very great, and thousands are doubtless passing continually undetected, because their light is wholly overpowered by that of the sun. Of those that are known, perhaps one in twelve develops into a naked-eye comet, and in some years six or seven will be discovered. With sufficiently powerful telescopes, there are as a rule not many weeks in the year when no comet is visible. Brilliant naked-eye comets are, however, infrequent.

Comets, except Halley's, generally bear the name of their discoverer, as Donati (1858), and Pons-Brooks (1893). Pons was a very active discoverer of comets in France early in the nineteenth century: he was a doorkeeper at the observatory of Marseilles, and his name is now more famous in astronomy than that of Thulis, then the director of the Observatory, who taught and encouraged him. Messier was another very successful discoverer of comets in France, and in America we have had many: Swift, Brooks, and Barnard the most successful.

How bright a comet will be and how long it will be visible depends upon many conditions. So the comets vary much in these respects. The first comet of 1811 was under observation for nearly a year and a half, the longest on record till Halley's in 1910. In case a comet eludes discovery and observation until it has passed its perihelion, or nearest point to the sun, its period of visibility may be reduced to a few weeks only. The brightest comets on record were visible in 1843 and 1882: so brilliant were they that even the effulgence of full daylight did not overpower them. In particular the comet of 1843 was not only excessively bright, but at its nearest approach to the earth its tail swept all the way across the sky from one horizon to the other. It must have looked very much like the straight beam of an enormous searchlight, though very much brighter.

The tails of comets are to the naked eye the most compelling thing about them, and to the ancient peoples they were naturally most terrifying. Their tails are not only curved, but sometimes curved with varying degrees of curvature, and this circumstance adds to their weirdness of appearance. If we examine the tail of a comet with a telescope, it vanishes as if there were nothing to it: as indeed one may almost say there is not. Ordinarily, only the head of the comet is of interest in the telescope. When first seen there is usually nothing but the head visible, and that is made up of portions which develop more or less rapidly, presenting a succession of phenomena quite different in different comets.

When first discovered a comet is usually at a great distance from the sun, about the distance of Jupiter; and we see it, not as we do the planets, by sunlight reflected from them, but by the comet's own light. This is at that time very faint, and nearly all comets at such a distance look alike: small roundish hazy patches of faint, cloudlike light, with very often a concentration toward the center called the nucleus, on the average about 4,000 miles in diameter. Approach toward the sun brightens up the comet more and more, and the nucleus usually becomes very much brighter and more starlike. Then on the sunward side of the nucleus, jetlike streamers or envelopes appear to be thrown off, often as if in parallel curved strata, or concentrically. As they expand and move outward from the nucleus, these envelopes grow fainter and are finally merged in the general nebulosity known as the comet's head, which is anywhere from 30,000 to 100,000 miles in diameter. As a rule, this is an orderly development which can be watched in the telescope from hour to hour and from night to night; but occasionally a cometary visitor is quite a law to itself in development, presenting a fascinating succession of unpredictable surprises.

Then follows the development of the comet's tail, perhaps more striking than anything that has preceded it. Here a genuine repulsion from the sun appears to come into play. It may be an electrical repulsion. Much of the material projected from the comet's nucleus, seems to be driven backward or repelled by the sun, and it is this that goes to form the tail. The particles which form the tail then travel in modified paths which nevertheless can be calculated. The tail is made up of these luminous particles and it expands in space much in the form of a hollow, horn-shaped cone, the nucleus being near the tip of the horn.

Some comets possess multiple tails with different degrees of curvature, Donati's for example. Usually there is a nearly straight central dark space, marking the axis of the comet, and following the nucleus. But occasionally this is replaced by a thin light streak very much less in breadth than the diameter of the head. Cometary tails are sometimes 100 million miles in length.

Three different types of cometary tails are recognized. First, the long straight ones, apparently made up of matter repelled by the sun twelve to fifteen times more powerfully than gravitation attracts it. Such particles must be brushed away from the comet's head with a velocity of perhaps five miles a second, and their speed is continually increasing. Probably these straight tails are due to hydrogen. The second type tails are somewhat curved, or plume-like, and they form the most common type of cometary tail. In them the sun's repulsion is perhaps twice its gravitational attraction, and hydrocarbons in some form appear to be responsible for tails of this character. Then there is a third type, much less often seen, short and quickly curving, probably due to heavier vapors, as of chlorine, or iron, or sodium, in which the repulsive force is only a small fraction of that of gravitation.

Many features of this theory of cometary tails are borne out by examination of their light with the spectroscope, although the investigation is as yet fragmentary. It is evident that the tail of a comet is formed at the expense of the substance of the nucleus and head; so that the matter repelled is forever dissipated through the regions of space which the comet has traveled. Comets must lose much of their original substance every time they return to perihelion. Comets actually age, therefore, and grow less and less in magnitude of material as well as brightness, until they are at last opaque, nonluminous bodies which it becomes impossible to follow with the telescope.

CHAPTER XLI
WHERE DO COMETS COME FROM?

Where do comets come from? The answer to this question is not yet fully made out. Most likely they have not all had a similar origin, and theories are abundant. Apparently they come into the solar system from outer space, from any direction whatsoever. The depths of interstellar space seem to be responsible for most, if not all, of the new ones. Whether they have come from other stars or stellar systems we cannot say.

While comets are tremendous in size or volume, their mass or the amount of real substance in them is relatively very slight. We know this by the effect they produce on planets that they pass near, or rather by the effect that they fail to produce. The earth's atmosphere weighs about one two hundred and fifty thousandth as much as the earth itself, but a comet's entire mass must be vastly less than this. Even if a comet were to collide with the earth head on, there is little reason to believe that dire catastrophe would ensue. At least twice the earth is known to have passed through the tail of a comet, and the only effect noticed was upon the comet itself; its orbit had been modified somewhat by the attraction of the earth. If the comet were a small one, collision with any of the planets would result in absorption and dissipation of the comet into vapor.

The whole of a large comet has perhaps as much mass or weight as a sphere of iron a hundred miles in diameter. Even this could not wreck the earth, but the effect would depend upon what part of the earth was hit. A comet is very thin and tenuous, because its relatively small mass is distributed through a volume so enormous. So it is probable that the earth's atmosphere could scatter and burn up the invading comet, and we should have only a shower of meteors on an unprecedented scale. Diffusion of noxious gases through the atmosphere might vitiate it to some extent, though probably not enough to cause the extinction of animal life.

Every comet has an interesting history of its own, almost indeed unique. One of the smallest comets and the briefest in its period round the sun is known as Encke's comet. It is a telescopic comet with a very short tail, its time of revolution is about three and a half years, and it exhibits a remarkable contraction of volume on approach to the sun.

Biela's comet has a period about twice as long. At one time it passed within about 15 million miles of the earth, and somewhere about the year 1840 this comet divided into two distinct comets, which traveled for months side by side, but later separated and both have since completely disappeared. Perhaps the most beautiful of all comets is that discovered by Donati of Florence in 1858. Its coma presented the development of jets and envelopes in remarkable perfection, and its tail was of the secondary or hydrocarbon type, but accompanied by two faint streamer tails, nearly tangential to the main tail and of the hydrogen type. Donati's comet moves in an ellipse of extraordinary length, and it will not return to the sun for nearly 2,000 years.

The most brilliant comet of the last half century is known as the great comet of 1882. In a clear sky it could readily be seen at midday. On September 17 it passed across the disk of the sun and was practically as bright as the surface of the sun itself. The comet had a multiple nucleus and a hydrocarbon tail of the second type, nearly a hundred million miles in length. Doubtless this great comet is a member of what is known as a cometary group, which consists of comets having the same orbit and traveling tandem round the sun. The comets of 1668, 1843, 1880, 1882 and 1887 belong to this particular group, and they all pass within 300,000 miles of the sun's surface, at a maximum velocity exceeding 300 miles a second. They must therefore invade the regions of the solar corona, the inference being that the corona as well as the comet is composed of exceedingly rare matter.

Photography of comets has developed remarkably within recent years, especially under the deft manipulation of Barnard, whose plates, in particular during his residence at the Lick Observatory on Mount Hamilton, California, show the features of cometary heads and tails in excellent definition. Halley's comet, at the 1910 apparition, was particularly well photographed at many observatories.

The question is often asked, When will the next comet come? If a large bright comet is meant, astronomers cannot tell. At almost any time one may blaze into prominence within only a few days. During the latter half of the last century, bright comets appeared at perihelion at intervals of eight years on the average. Several of the lesser and fainter periodic comets return nearly every year, but they are mostly telescopic, and are rarely seen except by astronomers who are particularly interested in observing them.

CHAPTER XLII

METEORS AND SHOOTING STARS

"Falling stars," or "shooting stars," have been familiar sights in all ages of the world, but the ancient philosophers thought them scarcely worthy of notice. According to Aristotle they were mere nothings of the upper atmosphere, of no more account than the general happenings of the weather. But about the end of the eighteenth century and the beginning of the nineteenth the insufficiency of this view began to be fully recognized, and interplanetary space was conceived as tenanted by shoals of moving bodies exceedingly small in mass and dimension as compared with the planets.

Millions of these bodies are all the time in collision with the outlying regions of our atmosphere; and by their impact upon it and their friction in passing swiftly through it, they become heated to incandescence, thus creating the luminous appearances commonly known as shooting stars. For the most part they are consumed or dissipated in vapor before reaching the solid surface of the earth; but occasionally a luminous cloud or streak is left glowing in the wake of a large meteor, which sometimes remains visible for half an hour after the passage of the meteor itself. These mistlike clouds projected upon the dark sky have been especially studied by Trowbridge of Columbia University.

Many more meteors are seen during the morning hours, say from four to six, than at any other nightly period of equal length, because the visible sky is at that time nearly centered around the general direction toward which the earth is moving in its orbit round the sun; so that the number of meteors that would fall upon the earth if at rest is increased by those which the earth overtakes by its own motion. Also from January to July while the earth is traveling from perihelion to aphelion, fewer meteors are seen than in the last half of the year; but this is chiefly because of the rich showers encountered in August and November.

Although the descent of meteoric bodies from the sky was pretty generally discredited until early in the nineteenth century, such falls had nevertheless been recorded from very early times. They were usually regarded as prodigies or miracles, and such stones were commonly objects of worship among ancient peoples. For example, the Phrygian Stone, known as the "Diana of the Ephesians which fell down from Jupiter," was a famous stone built into the Kaaba at Mecca, and even to-day it is revered by Mohammedans as a holy relic. Perhaps the earliest known meteoric fall is that historically recorded in the Parian Chronicle as having occurred in the island of Crete, B. C. 1478. Also in the imperial museum of Petrograd is the Pallas or Krasnoiarsk iron, perhaps three-quarters of a ton in weight, found in 1772 by Pallas, the famous traveler, at Krasnoiarsk, Siberia.

But a fall of meteoric stones that chanced upon the department of Orne, France, in 1805, led to a critical investigation by Biot, the distinguished physicist and academician. According to his report a violent explosion in the neighborhood of L'Aigle had been heard for a distance of seventy-five miles around, and lasting five or six minutes, about 1 P. M. on Tuesday, April 26. From several adjoining towns a rapidly moving fireball had been seen in a sky generally clear, and there was absolutely no room for doubt that on the same day many stones fell in the neighborhood of L'Aigle. Biot estimated their number between two and three thousand, and they were scattered over an elliptical area more than six miles long, and two and a half miles broad. Thenceforward the descent of meteoric matter from outer space upon the earth has been recognized as an unquestioned fact.

The origin of these bodies being cosmic, meteors may be expected to fall upon the earth without reference to latitude, or season, or day and night, or weather. On entering our upper atmosphere their temperature must be that of space, many hundred degrees below zero; and their velocities range from ten miles per second upward. But atmospheric resistance to their flight is so great that their velocity is quickly reduced: at ground impact it does not exceed a few hundred feet per second. On January 1, 1869, several meteoric stones fell on ice only a few inches thick in Sweden, rebounding without either breaking through the ice or being themselves fractured.

Naturally the flight of a meteor through the atmosphere will be only a few seconds in duration, and owing to the sudden reduction of velocity, it will continue to be luminous throughout only the upper part of its course. Visibility generally begins at an elevation of about seventy miles, and ends at perhaps half that altitude.

What is the origin of meteors? Theories there are in great abundance: that they come from the sun, that they come from the moon, that they come from the earth in past ages as a result of volcanic action, and so on. But there are many difficulties in the way of acceptance of these and several other theories. That all meteors were originally parts of cometary masses is however a theory that may be accepted without much hesitation.

Comets have been known to disintegrate. Biela's comet even disappeared entirely, so that during a shower of Biela meteors in November, 1885, an actual fragment of the lost comet fell upon the earth, at Mazapil, Mexico. And as the Bielid meteors encounter the earth with the relatively low velocity of ten miles a second, we may expect to capture other fragments in the future. Numerous observers saw the weird disintegration of the nucleus of the great comet of 1882, well recognized as a member of the family of the comet of 1843. As these comets are fellow voyagers through space along the same orbit, probably all five members of the family, with perhaps others, were originally a single comet of unparalleled magnitude.

The Brooks comet of 1890 affords another instance of fragmentary nucleus. The oft-repeated action of solar forces tending to disrupt the mass of a comet more and more, and scatter its material throughout space, the secular dismemberment of all comets becomes an obvious conclusion. During the hundreds of millions of years that these forces are known to have been operant, the original comets have been broken up in great numbers, so that elliptical rings of opaque meteoric bodies now travel round the sun in place of the comets.

These bodies in vast numbers are everywhere through space, each too small to reflect an appreciable amount of sunlight, and becoming visible only when they come into collision with our outer atmosphere. The practical identity of several such meteor streams and cometary orbits has already been established, and there is every reason for assigning a similar origin to all meteoric bodies. Meteors, then, were originally parts of comets, which have trailed themselves out to such extent that particles of the primal masses are liable to be picked up anywhere along the original cometary paths. The historic records of all countries contain trustworthy accounts of meteoric showers. Making due allowances for the flowery imagery of the oriental, it is evident that all have at one time or another seen much the same thing. In A. D. 472, for instance, the Constantinople sky was reported alive with flying stars. In October, 1202, "stars appeared like waves upon the sky; and they flew about like grasshoppers." During the reign of King William II occurred a very remarkable shower in which "stars seemed to fall like rain from heaven."

But the showers of November, 1799 and 1833, are easily the most striking of all. The sky was filled with innumerable fiery trails and there was not a space in the heavens a few times the size of the moon that was not ablaze with celestial fireworks. Frequently huge meteors blended their dazzling brilliancy with the long and seemingly phosphorescent trails of the shooting stars.

The interval of thirty-four years between 1799 and 1833 appeared to indicate the possibility of a return of the shower in November of 1866 or 1867, and all the people of that day were aroused on this subject and made every preparation to witness the spectacle. Extemporized observatories were established, watchmen were everywhere on the lookout, and bells were to be rung the minute the shower began. The newspapers of the day did little to allay the fears of the multitude, but the critical days of November, 1866, passed with disappointment in America. In Europe, however, a fine shower was seen, though it was not equal to that of 1833. The astronomers at Greenwich counted many thousand meteors. In November of 1867, however, American astronomers were gratified by a grand display, which, although failing to match the general expectation, nevertheless was a most striking spectacle, and the careful preparation for observing it afforded data of observation which were of the greatest scientific value. The actual orbits of these bodies in space became known with great exactitude, and it was found that their general path was identical with that of the first comet of 1866, which travels outward somewhat beyond the planet Uranus. When the visible paths of these meteors are traced backward, all appear as if they originated from the constellation Leo. So they are known as Leonids, and a return of the shower was confidently predicted for November, 1900-1901, which for unknown reasons failed to appear.

Two Views of Halley's Comet. Taken with the same camera from the same position, one on May 12, and the other on May 15, 1910.

(Photo, Mt. Wilson Solar Observatory.)

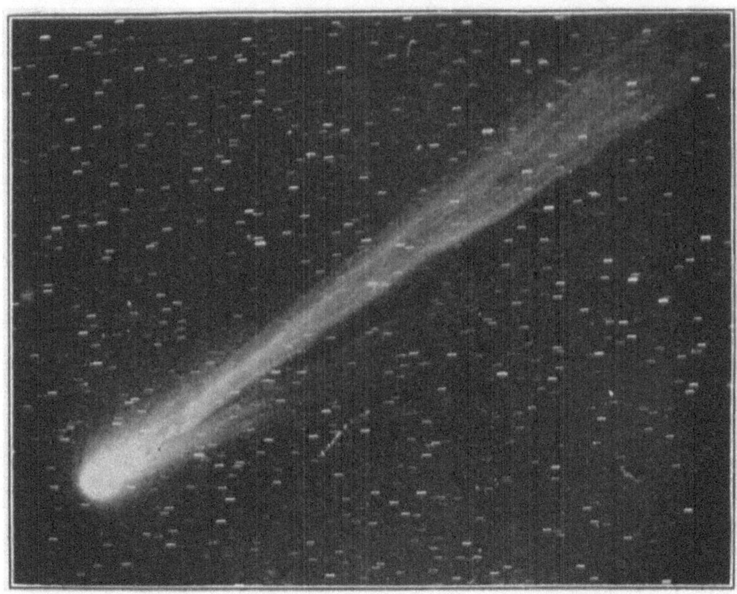

Swift's Comet of 1892. This comet showed extraordinary and rapid transformations, one day having a dozen streamers in its tail, another only two.

(Photo by Prof. E. E. Barnard.)

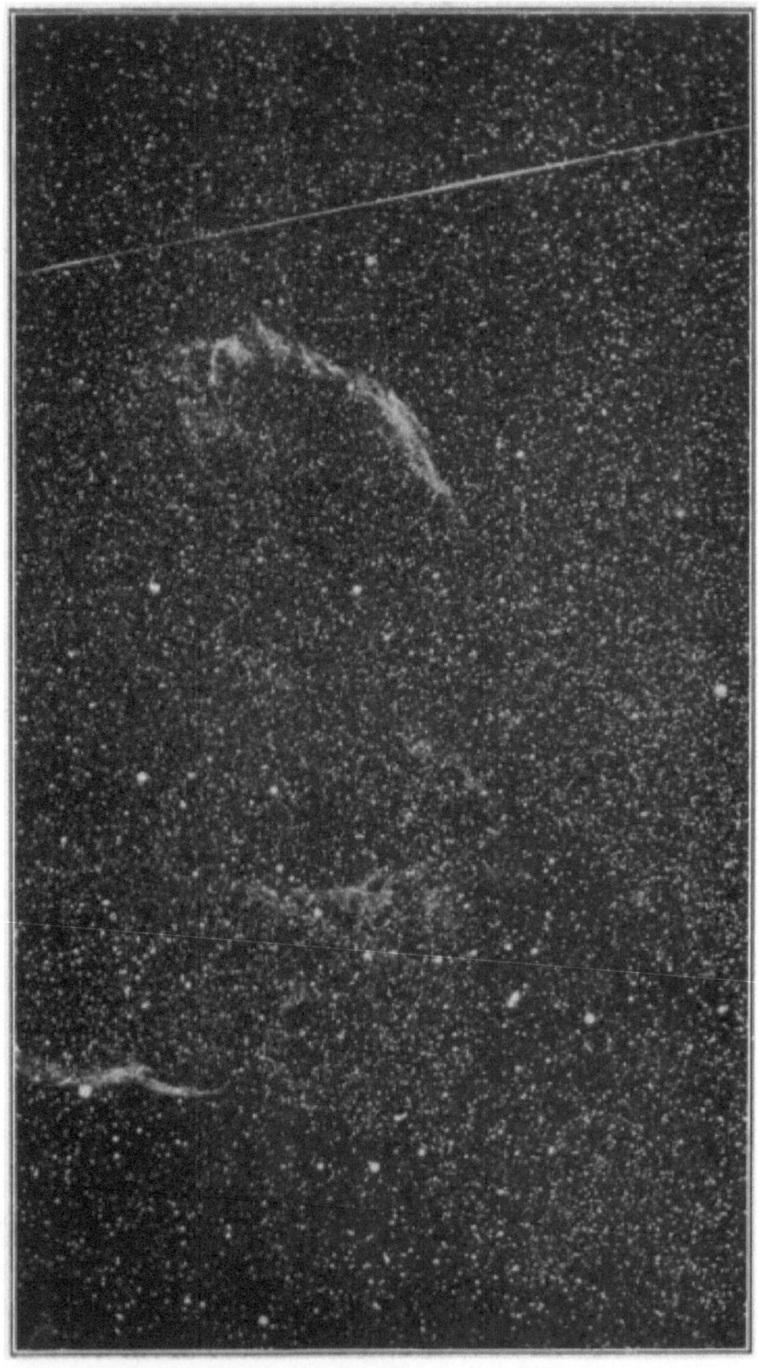

A Large Meteor Trail In The Field With Fine Nebulæ.

(Photo, Yerkes Observatory.)

During the last half century meteors have been pretty systematically observed, especially by the astronomers of Italy and Denning of England, so that several hundred distinct showers are now known, their radiant points fall in every part of the heavens, and there is scarcely a clear moonless night when careful watching for meteors will be unrewarded. Besides November, the months of August (Perseids), April (Lyrids), and December (Geminids) are favorable. Following in tabular form is a fairly comprehensive list of the meteoric showers of the year, with the positions of the radiant points and the epochs of the showers according to Denning:

RADIANT POINT			
Name of Shower	**R. A.**	**Decl.**	**Date of Shower**
Quadrantids	230°	+53°	Jan. 2-4
Zeta Cepheids	331°	+56°	Jan. 25
Alpha Leonids	155°	+14°	Feb. 19-March 1
Tau Leonids	166°	+4°	March 1-4
Beta Ursids	161°	+58°	March 13-24
Lyrids	271°	+33°	April 20-22
Gamma Aquarids	338°	-2°	May 1-6
Zeta Herculids	246°	+29°	May 18-26
Eta Pegasids	330°	+28°	May 30-June 4
Theta Boötids	213°	+53°	June 27-28
Alpha Capricornids	304°	-12°	July 15-28
Delta Aquarids	339°	-11°	July 25-30
Perseids	45°	+57°	Aug. 10-12
Omicron Draconids	291°	+60°	Aug. 15-25
Zeta Draconids	262°	+63°	Aug. 21-Sept. 2
Piscids	348°	+2°	Sept. 4-14
Alpha Andromedids	4°	+28°	Sept. 27
Epsilon Arietids	40°	+20°	Oct. 11-24
Orionids	92°	+15°	Oct. 17-24
Epsilon Perseids	61°	+35°	Nov. 5
Leonids	150°	+23°	Nov. 13-15
Epsilon Taurids	64°	+22°	Nov. 14-25
Andromedids	25°	+43°	Nov. 17-23
Beta Geminids	119°	+31°	Dec. 1-12
Geminids	108°	+33°	Dec. 1-14
Alpha Ursæ Majorids	161°	+58°	Dec. 18-21
Kappa Draconids	194°	+68°	Dec. 18-28

The year 1916 was exceptional in providing an abundant and previously unknown shower on June 28, and its stream has nearly the same orbit as that of the Pons-Winnecke periodic comet. Useful observations of meteors are not difficult to make, and they are of service to professional astronomers investigating the orbits of these bodies, among whom are Mitchell and Olivier of the University of Virginia.

CHAPTER XLIII
METEORITES

Meteorites, the name for meteors which have actually gone all the way through our atmosphere, are never regular in form or spherical. As a rule the iron meteorites are covered with pittings or thumb marks, due probably to the resistance and impact of the little columns of air which impede its progress, together with the unequal condition and fusibility of their surface material. The work done by the atmosphere in suddenly checking the meteor's velocity appears in considerable part as heat, fusing the exterior to incandescence. This thin liquid shell is quickly brushed off, making oftentimes a luminous train.

But notwithstanding the exceedingly high temperature of the exterior, enforced upon it for the brief time of transit through the atmosphere, it is probable that all large meteorites, if they could be reached at once on striking the earth, would be found to be cold, because the smooth, black, varnishlike crust which always incases them as a result of intense heat is never thick. On one occasion a meteor which was seen to fall in India was dug out of the ground as quickly as possible, and found to be, not hot as was expected, but coated thickly over with ice frozen on it from the moisture in the surrounding soil.

As to the composition of shooting stars, and their probable mass, and its effect upon the earth, our data are quite insufficient. The lines of sodium and magnesium have been hurriedly caught in the spectroscope, and, estimating on the basis of the light emitted by them, the largest meteors must weigh ounces rather than pounds. Nevertheless, it is interesting to inquire what addition the continual fall of many millions daily upon the earth makes to its weight: somewhere between thirty and fifty thousand tons annually is perhaps a conservative estimate, but even this would not accumulate a layer one inch in thickness over the entire surface of the earth in less than a thousand million years.

Many hundreds of the meteors actually seen to fall, together with those picked up accidentally, are recovered and prized as specimens of great value in our collections, the richest of which are now in New York, Paris, and London. The detailed investigation of them is rather the province of the chemist, the crystallographer and the mineralogist than of the astronomer whose interest is more keen in their life history before they reach the earth. To distinguish a stony meteorite from terrestrial rock substances is not always easy, but there is usually little difficulty in pronouncing upon an iron meteorite. These are most frequently found in deserts, because the dryness of the climate renders their oxidation and gradual disappearance very slow.

The surface of a suspected iron meteorite is polished to a high luster and nitric acid is poured upon it. If it quickly becomes etched with a characteristic series of lines, or a sort of cross-hatching, it is almost certain to be a meteorite. Occasionally carbon has been found in meteorites, and the existence of diamond has been suspected. The minerals composing meteorites are not unlike terrestrial materials of volcanic origin, though many of them are peculiar to meteorites only. More than one-third of all the known chemical elements have been found by analysis in meteorites, but not any new ones.

Meteoric iron is a rich alloy containing about ten per cent of nickel, also cobalt, tin, and copper in much smaller amount. Calcium, chlorine, sodium, and sulphur likewise are found in meteoric irons. At very high temperatures iron will absorb gases and retain them until again heated to red heat. Carbonic oxide, helium, hydrogen, and nitrogen are thus imprisoned, or occluded, in meteoric irons in very small quantities; and in 1867, during a London lecture by Graham, a room in the Royal Institution was for a brief space illuminated by gas brought to earth in a meteorite from interplanetary space. Meteorites, too, have been most critically investigated by the biologist, but no trace of germs of organic life of any type has so far been found. Farrington of Chicago has published a full descriptive catalogue of all the North American meteorites.

Recent investigations of the radioactivity of meteorites show that the average stone meteorite is much less radioactive than the average rock, and probably less than one-fourth as radioactive as in average granite. The metallic meteorites examined were found about wholly free from radioactivity.

From shooting stars, perhaps the chips of the celestial workshop, or more possibly related to the planetesimals which the processes of growth of the universe have swept up into the vastly greater bodies of the universe, transition is natural to the stars themselves, the most numerous of the heavenly bodies, all shining by their own light, and all inconceivably remote from the solar system, which nevertheless appears to be not far removed from the center of the stellar universe.

CHAPTER XLIV

THE UNIVERSE OF STARS

Our consideration of the solar system hitherto has kept us quite at home in the universe. The outer known planets, Uranus and Neptune, are indeed far removed from the sun, and a few of the comets that belong to our family travel to even greater distances before they begin to retrace their steps sunward. When we come to consider the vast majority of the glistening points on the celestial sphere—all in fact except the five great planets, Mercury, Venus, Mars, Jupiter, and Saturn—we are dealing with bodies that are self-luminous like the sun, but that vary in size quite as the bodies of the solar system do, some stars being smaller than the sun and others many hundred fold larger than he is; some being "giants," and others "dwarfs." But the overwhelming remoteness of all these bodies arrests our attention and even taxes our credulity regarding the methods that astronomers have depended on to ascertain their distances from us.

Their seeming countlessness, too, is as bewildering as are the distances; though, if we make actual counts of those visible to the naked eye within a certain area, in the body of the "Great Bear," for example, the great surprise will be that there are so few. And if the entire dome of the sky is counted, at any one time, a clear, moonless sky would reveal perhaps 2,500, so that in the entire sky, northern and southern, we might expect to find 5,000 to 6,000 lucid stars, or stars visible to the naked eye.

But when the telescope is applied, every accession of power increases the myriads of fainter and fainter stars, until the number within optical reach of present instruments is somewhere between 400 and 500 millions. But if we were to push the 100-inch reflector on Mount Wilson to its limit by photography with plates of the highest sensitiveness, millions upon millions of excessively faint stars would be plainly visible on the plates which the human eye can never hope to see directly with any telescope present or future, and which would doubtless swell the total number of stars to a thousand millions. Recent counts of stars by Chapman and Melotte of Greenwich tend to substantiate this estimate.

What have astronomers done to classify or catalogue this vast array of bodies in the sky? Even before making any attempt to estimate their number, there is a system of classification simply by the amount of light they send us, or by their apparent stellar magnitudes—not their actual magnitudes, for of those we know as yet very little. We speak of stars of the "first magnitude," of which there are about 20, Sirius being the brightest and Regulus the faintest. Then there are about 65 of the second, or next fainter, magnitude, stars like Polaris, for example, which give an amount of light two and a half times less than the average first magnitude star. Stars of the third magnitude are fainter than those of the second in the same ratio, but their number increases to 200; fourth magnitude, 500; fifth magnitude, 1,400; sixth magnitude, 5,000, and these are so faint that they are just visible on the best nights without telescopic aid.

Decimals express all intermediate graduations of magnitude. Astronomers carry the telescopic magnitudes much farther, till a magnitude beyond the twentieth is reached, preserving in every case the ratio of two and one-half for each magnitude in relation to that numerically next to it. Even Jupiter and Venus, and the sun and moon, are sometimes calculated on this scale of stellar magnitude, numerically negative, of course, Venus sometimes being as bright as magnitude -4.3, and the sun -26.7.

Knowing thus the relation of sun, moon, and stars, and the number of the stars of different magnitudes, it is possible to estimate the total light from the stars. This interesting relation comes out this way: that the stars we cannot see with the naked eye give a greater total of light than those we can because of their vastly greater numbers. And if we calculate the total light of all the brighter stars down to magnitude nine and one-half, we find it equal to $\frac{1}{80}$th of the light of the average full moon.

Many stars show marked differences in color, and strictly speaking the stars are now classified by their colors. The atmosphere affects star colors very considerably, low altitudes, or greater thickness of air, absorbing the bluish rays more strongly and making the stars appear redder than they really are. Aldebaran, Betelgeuse and Antares are well-known red stars, Capella and Alpha Ceti yellowish, Vega and Sirius blue, and Procyon and Polaris white. Among the telescopic stars are many of a deep blood-red tint, variable stars being numerous among them. Double stars, too, are often complementary in color. There is evidence indicating change of color of a very few stars in long periods of time; Sirius, for example, two thousand years ago was a red star, now it is blue or bluish white. But the meaning of color, or change of color in a star is as yet only incompletely ascertained. It may be connected with the radiative intensity of the star, or its age, or both.

The late Professor Edward C. Pickering was famous for his life-long study and determination of the magnitudes of the stars. Standards of comparison have been many, and have led to much unnecessary work. Pickering chose Polaris as a standard and devised the meridian photometer, an ingenious instrument of high accuracy, in which the light of a star is compared directly with that of the pole star by reflection. All the bright stars of both the northern and the southern skies are worked into a standard system of magnitudes known as HP, or the Harvard Photometry.

Astronomers make use of several different kinds of magnitude for the stars: the apparent magnitude, as the eye sees it, often called the visual magnitude; the photographic magnitude, as the photographic plate records it, and these are now determined with the highest accuracy; the photovisual magnitude, quite the same as the visual, but determined photographically on an isochromatic plate with a yellow screen or filter, so that the intensity is nearly the same as it appears to the eye. The difference between the star's visual or photovisual magnitude and its photographic magnitude is called its color-index, and is often used as a measure of the star's color. Light of the shorter wave lengths, as blue and violet, affects the photographic plate more rapidly than the reds and yellows of longer wave length by which the eye mainly sees; so that red stars will appear much fainter and blue stars much brighter on the ordinary photographic plate than the eye sees them.

So great are the differences of color in the stars that well-known asterisms, with which the eye is perfectly familiar, are sometimes quite unrecognizable on the photographic plate, except by relative positions of the stars composing them. White stars affect the eye and the plate about equally, so that their visual or photovisual and photographic magnitudes are about equal. The studies of the colors of the stars, the different methods of determining them, and the relations of color to constitution have been made the subject of especial investigation by Seares of Mount Wilson and many other astronomers.

Centuries of the work of astronomers have been faithfully devoted to mapping or charting the stars and cataloguing them. Just as we have geographical maps of countries, so the heavens are parceled out in sections, and the stars set down in their true relative positions just as cities are on the map. Recent years have added photographic charts, especially of detailed regions of the sky; but owing to spectral differences of the stars, their photographic magnitudes are often quite different from their visual magnitudes. From these maps and charts the positions of the stars can be found with much precision; but if we want the utmost accuracy, we must go to the star catalogues—huge volumes oftentimes, with stellar positions set down therein with the last degree of precision.

First there will be the star's name, and in the next column its magnitude, and in a third the star's right ascension. This is its angular distance eastward around the celestial sphere starting from the vernal equinox, and it corresponds quite closely to the longitude of a place which we should get from a gazetteer, if we wished to locate it on the earth. Then another column of the catalogue will give the star's declination, north or south of the equator, just as the gazetteer will locate a city by its north or south latitude.

CHAPTER XLV
STAR CHARTS AND CATALOGUES

Who made the first star chart or catalogue? There is little doubt that Eudoxus (B. C. 200) was the first to set down the positions of all the brighter stars on a celestial globe, and he did this from observations with a gnomon and an armillary sphere. Later Hipparchus (B. C. 130) constructed the first known catalogue of stars, so that astronomers of a later day might discover what changes are in progress among the stars, either in their relative positions or caused by old stars disappearing or new stars appearing at times in the heavens. Hipparchus was an accurate observer, and he discovered an apparent and perpetual shifting of the vernal equinox westward, by which the right ascensions of the stars are all the time increasing. He determined the amount of it pretty accurately, too. His catalogue contained 1,080 stars, and is printed in the "Almagest" of Ptolemy.

Centuries elapsed before a second star catalogue was made, by Ulugh-Beg, an Arabian astronomer, A. D. 1420, who was a son of Tamerlane, the Tartar monarch of Samarcand, where the observations for the catalogue were made. The stars were mainly those of Ptolemy, and much the same stars were reobserved by Tycho Brahe (A. D. 1580) with his greatly improved instruments, thus forming the third and last star catalogue of importance before the invention of the telescope.

From the end of the seventeenth century onward, the application of the telescope to all the types of instruments for making observations of star places has increased the accuracy many-fold. The entire heavens has been covered by Argelander in the northern hemisphere, and Gould in the southern—over 700,000 stars in all. Many government observatories are still at work cataloguing the stars. The Carnegie Institution of Washington maintains a department of astrometry under Boss of Albany, which has already issued a preliminary catalogue of more than 6,000 stars, and has a great general catalogue in progress, together with investigations of stellar motions and parallaxes. This catalogue of star positions will include proper motions of stars to the seventh magnitude.

In 1887 on proposal of the late Sir David Gill, an international congress of astronomers met at Paris and arranged for the construction of a photographic chart of the entire heavens, allotting the work to eighteen observatories, equipped with photographic telescopes essentially alike. The total number of plates exceeds 25,000. Stars of the fourteenth magnitude are recorded, but only those including the eleventh magnitude will be catalogued, perhaps 2,000,000 in all. The expense of this comprehensive map of the stars has already exceeded \$2,000,000, and the work is now nearly complete. Turner of Oxford has conducted many special investigations that have greatly enhanced the progress of this international enterprise.

Other great photographic star charts have been carried through by the Harvard Observatory, with the annex at Arequipa, Peru, employing the Bruce photographic telescope, a doublet with 24-inch lenses; also Kapteyn of Groningen has catalogued about 300,000 stars on plates taken at Cape Town. Charting and cataloguing the stars, both visually and photographically, is a work that will never be entirely finished. Improvements in processes will be such that it can be better done in the future than it is now, and the detection of changes in the fainter stars and investigation of their motions will necessitate repetition of the entire work from century to century.

The origin of the names of individual stars is a question of much interest. The constellation figures form the basis of the method, and the earliest names were given according to location in the especial figure; as for instance, Cor Scorpii, the heart of the Scorpion, later known as Antares or Alpha Scorpii. The Arabians adopted many star names from the Greeks, and gave about a hundred special names to other stars. Some of these are in common use to-day, by navigators, observers of meteors and of variable stars. Sirius, Vega, Arcturus, and a few other first magnitude stars, are instances.

But this method is quite insufficient for the fainter stars whose numbers increase so rapidly. Bayer, a contemporary of Galileo, originated our present system, which also employs the names of the constellations, the Latin genitive in each case, prefixed by the small letters of the Greek alphabet, from alpha to omega, in order of decreasing brightness; and followed by the Roman letters when the Greek alphabet is exhausted.

If there were still stars left in a constellation unnamed, numbers were used, first by Flamsteed, Astronomer Royal; and numbers in the order of right ascension in various catalogues are used to designate hundreds of other stars. The vast bulk of the stars are, however, nameless; but about one million are identifiable by their positions (right ascension and declination) on the celestial sphere.

CHAPTER XLVI
THE SUN'S MOTION TOWARD LYRA

If Hipparchus or Galileo should return to earth to-night and look at the stars and constellations as we see them, there would be no change whatever discernible in either the brightness of the stars or in their relative positions. So the name fixed stars would appear to have been well chosen. Halley in the seventeenth century was the first to detect that slow relative change of position of a few stars which is known as proper motion, and all the modern catalogues give the proper motions in both right ascension and declination. These are simply the small annual changes in position athwart the line of vision; and, as a whole, the proper motions of the brighter stars exceed the corresponding motions of the fainter ones because they are nearer to us. The average proper motion of the brightest stars is 0".25, and of stars of the sixth magnitude only one-sixth as great.

A few extreme cases of proper motion have been detected, one as large as 9", of an orange yellow star of the eighth magnitude in the southern constellation Pictor, and Barnard has recently discovered a star with a proper motion exceeding 10"; several determinations of its parallax give 0".52, corresponding to a distance of 6.27 light years. Nevertheless, two centuries would elapse before these stars would be displaced as much as the breadth of the moon among their neighbors in the sky. The proper motions of stars are along perfectly straight lines, so far as yet observed. Ultimately we may find a few moving in curved paths or orbits, but this is hardly likely.

As for a central sun hypothesis, that pointing out Alcyone in particular, there is no reliable evidence whatever. Analysis of the proper motions of stars in considerable numbers, first by Sir William Herschel, showed that they were moving radially from the constellation Hercules, and in great numbers also toward the opposite side of the stellar sphere. Later investigation places this point, called the sun's goal, or apex of the sun's way, over in the adjacent constellation Lyra; and the opposite point, or the sun's quit, is about halfway between Sirius and Canopus. By means of the radial velocities of stars in these antipodal regions of the sky, it is found that the sun's motion toward Lyra, carrying all his planetary family along with him, is taking place at the rate of about 12 miles in every second.

While the right ascensions of the solar apex as given by the different investigations have been pretty uniform, the declination of this point has shown a rather wide variation not yet explained. For example, there is a difference of nearly ten degrees between the declination (+34°.3) of the apex as determined by Boss from the proper motions of more than 6,000 stars, and the declination (+25°.3) found by Campbell from the radial velocities of nearly 1,200 stars. Several investigations tend to show that the fainter the stars are, the greater is the declination of the solar apex. More remarkable is the evidence that this declination varies with the spectral type of the stars, the later types, especially G and K, giving much more northerly values. On the whole the great amount of research that has been devoted to the solar motion relative to the system of the stars for the past hundred years may be said to indicate a point in right ascension 18h. (270°) and declination 34° N. as the direction toward which the sun is moving. This is not very far from the bright star Alpha Lyræ, and the antipodal point from which the sun is traveling is quite near to Beta Columbæ.

So swift is this motion (nearly twenty kilometers per second) that it has provided a base line of exceptional length, and very great service in determining the average distance of stars in groups or classes. After thousands of years the sun's own motion combined with the proper motions of the stars will displace many stars appreciably from their familiar places. The constellations as we know them will suffer slight distortions, particularly Orion, Cassiopeia and Ursa Major. Identity or otherwise of spectra often indicates what stars are associated together in groups, and their community of motion is known as star drift. Recent investigation of vast numbers of stars by both these methods have led to the epochal discovery of star streaming, which indicates that the stars of our system are drifting by, or rather through, each other, in two stately and interpenetrating streams. The grand primary cause underlying this motion is as yet only surmised.

CHAPTER XLVII
STARS AND THEIR SPECTRAL TYPE

When in 1872 Dr. Henry Draper placed a very small wet plate in the camera of his spectroscope and, by careful following, on account of the necessarily long exposure, secured the first photographic spectrum of a star ever taken, he could hardly have anticipated the wealth of the new field of research which he was opening. His wife, Anna Palmer Draper, was his enthusiastic assistant in both laboratory and observatory, and on his death in 1882, she began to devote her resources very considerably to the amplification of stellar spectrum photography. At first with the cooperation of Professor Young of Princeton, and later through extension of the facilities of Harvard College Observatory, whose director, the late Professor Edward C. Pickering, devoted his energies in very large part to this matter, all the preliminaries of the great enterprise were worked out, and a comprehensive program was embarked upon, which culminated in the "Henry Draper Memorial," a catalogue and classification of the spectra of all the stars brighter than the ninth magnitude, in both the northern and southern hemispheres.

One very remarkable result from the investigation of large numbers of stars according to their type is the close correlation between a star's luminosity and its spectral type. But even more remarkable is the connection between spectral type and speed of motion. As early as 1892 Monck of Dublin, later Kapteyn, and still later Dyson, directed attention to the fact that stars of the Secchi type II had on the average larger proper motions than those of type I. In 1903 Frost and Adams brought out the exceptional character of the Orion stars, the radial velocities of twenty of which averaged only seven kilometers per second.

Soon after, with the introduction of the two-stream hypothesis, a wider generalization was reached by Campbell and Kapteyn, whose radial velocities showed that the average linear velocity increases continually through the entire series B, A, F, G, K, M, from the earliest types of evolution to the latest. The younger stars of early type have velocities of perhaps five or six kilometers per second, while the older stars of later type have velocities nearly fourfold greater.

The great question that occurs at once is: How do the individual stars get their motions? The farther back we go in a star's life history, the smaller we find its velocity to be. When a star reaches the Orion stage of development, its velocity is only one-third of what it may be expected to have finally. Apparently, then, the stars at birth have no motion, but gradually acquire it in passing through their several types or stages of development.

More striking still is the motion of the planetary nebulæ, in excess of 25 kilometers per second, while type A stars move 11 kilometers, type G 15 kilometers, and type M 17 kilometers per second. Can the law connecting speed of motion and spectral type be so general that the planetary nebula is to be regarded as the final evolutionary stage? Stars have been seen to become nebulæ, and one astronomer at least is strongly of the opinion that a single such instance ought to outweigh all speculation to the contrary, as that stars originate from nebulæ.

In his discussion of stellar proper motions, Boss has reached a striking confirmation of the relation of speed to type, finding for the cross linear motion of the different types a series of velocities closely paralleling those of Kapteyn and Campbell.

Concerning the marked relation of the luminosities of the stars to their spectral types, there is a pronounced tendency toward equality of brightness among stars of a given type; also the brightness diminishes very markedly with advance in the stage of evolution. There has been much discussion as to the order of evolution as related to the type of spectrum, and Russell of Princeton has put forward the hypothesis of giant stars and dwarf stars, each spectral type having these two divisions, though not closely related. One class embraces intensely luminous stars, the other stars only feebly luminous. When a star is in process of contraction from a diffused gaseous mass, its temperature rises, according to Lane's law, until that density is reached where the loss of heat by radiation exceeds the rise in temperature due to conversion of gravitational energy into heat. Then the star begins to cool again. So that if the spectrum of a star depends mainly on the effective temperature of the body, clearly the classification of the Draper catalogue would group stars together which are nearly alike in temperature, taking no note as to whether their present temperature is rising or falling.

Another classification of stars by Lockyer divides them according to ascending and descending temperatures. Russell's theory would assign the succession of evolutionary types in the order, M_1, K_1, G_1, F_1, A_1, B, A_2, F_2, G_2, K_2, M_2, the subscript 1 referring to the "giants," and 2 to the dwarf stars. In large part the weight of evidence would appear to favor the order of the Harvard classification, independently confirmed as it is by studies of stellar velocities, Galactic distribution, and periods of binary stars both spectroscopic and visual, where Campbell and Aiken find a marked increase in length of period with advance in spectral type. At the same time, a vast amount of evidence is accumulating in support of Russell's theory. Investigations in progress will doubtless reveal the ground on which both may be harmonized.

The publication of the new Henry Draper Catalogue of Stellar Spectra is in progress, a work of vast magnitude. The great catalogue of thirty years ago embraced the spectra of more than ten thousand stars, and was a huge work for that day; but the new catalogue utterly dwarfs it, with a classification much more detailed than in the earlier work, and with the number of stars increased more than twenty-fold. This work, projected by the late director of the Harvard Observatory, has been brought to a conclusion by the energy and enthusiasm of Miss Annie J. Cannon through six years of close application, aided by many assistants. The catalogue ranges over the stars of both hemispheres, and is a monument to masterly organization and completed execution which will be of the highest importance and usefulness in all future researches on the bodies of the stellar universe.

CHAPTER XLVIII
STAR DISTANCES

So vast are the distances of the stars that all attempts of the early astronomers to ascertain them necessarily proved futile. This led many astronomers after Copernicus to reject his doctrine of the earth's motion round the sun, so that they clung rather to the Ptolemaic view that the earth was without motion and was the center about which all the celestial motions took place. The geometry of stellar distances was perfectly understood, and many were the attempts made to find the parallaxes and distances of the stars; but the art of instrument making had not yet advanced to a stage where astronomers had the mechanisms that were absolutely necessary to measure very small angles.

About 1835, Bessel undertook the work of determining stellar parallax in earnest. His instrument was the heliometer, originally designed for measuring the sun's diameter; but as modified for parallax work it is the most accurate of all angle-measuring instruments that the astronomers employ. The star that he selected was 61 Cygni, not a bright star, of the sixth magnitude only, but its large proper motion suggested that it might be one of those nearest to us. He measured with the heliometer, at opposite seasons of the year, the distance of 61 Cygni from another and very small star in the same field of view, and thus determined the relative parallax of the two stars. The assumption was made that the very faint star was very much more distant than the bright one, and this assumption will usually turn out to be sound. Bessel got 0".35 for his parallax of 61 Cygni, and Struve by applying the same method to Alpha Lyræ, about the same time, got 0".25 for the parallax of that star.

These classic researches of Bessel and Struve are the most important in the history of star distances, because they were the first to prove that stellar parallax, although minute, could nevertheless be actually measured. About the same time success was achieved in another quarter, and Henderson, the British astronomer at the Cape of Good Hope, found a parallax of nearly a whole second for the bright star Alpha Centauri.

Although the parallaxes of many hundreds of stars have been measured since, and the parallaxes of other thousands of stars estimated, the measured parallax of Alpha Centauri, as later investigated by Elkin and Sir David Gill, and found to be 0".75, is the largest known parallax, and therefore Alpha Centauri is our nearest neighbor among the stars, so far as we yet know. This star is a binary system and the light of the two components together is about the same as that of Capella (Alpha Aurigæ). But it is never visible from this part of the world, being in 60 degrees of south declination: one might just glimpse it near the southern horizon from Key West.

How the distances of the stars are found is not difficult to explain, although the method of doing it involves a good deal of complication, interesting to the practical astronomer only. Recall the method of getting the moon's distance from the earth: it was done by measuring her displacement among the stars as seen from two widely separated observatories, as near the ends of a diameter of the earth as convenient. This is the base line, and the angle which a radius of the earth as seen from the center of the moon fills, or subtends, is the moon's parallax.

So near is the moon that this angle is almost an entire degree, and therefore not at all difficult to measure. But if we go to the distance of even Alpha Centauri, the nearest of the stars, our earth shrinks to invisibility; so that we must seek a longer base line. Fortunately there is one, but although its length is 25,000 times the earth's diameter, it is only just long enough to make the star distances measurable. We found that the sun's distance from the earth was 93 million miles; the diameter of the earth's orbit is therefore double that amount. Now conceive the diameter of the earth replaced by the diameter of the earth's orbit: by our motion round the sun we are transported from one extremity of this diameter to the opposite one in six month's time; so we may measure the displacement of a star from these two extremities, and half this displacement will be the star's parallax, often called the annual parallax because a year is consumed in traversing its period. And it is this very minute angle which Bessel and Struve were the first to measure with certainty, and which Henderson found to be in the case of Alpha Centauri the largest yet known.

Evidently the earth by its motion round the sun makes every star describe, a little parallactic ellipse; the nearer the star is the larger this ellipse will be, and the farther the star the smaller: if the star were at an infinite distance, its ellipse would become a point, that is, if we imagine ourselves occupying the position of the star, even the vast orbit of the earth, 186 million miles across, would shrink to invisibility or become a mathematical point.

Measurement of stellar parallax is one of many problems of exceeding difficulty that confront the practical astronomer. But the actual research nowadays is greatly simplified by photography, which enables the astronomer to select times when the air is not only clear, but very steady for making the exposures. Development and measurement of the plates can then be done at any time. Pritchard of Oxford, England, was among the earliest to appreciate the advantages of photography in parallax work, and Schlesinger, Mitchell, Miller, Slocum and Van Maanen, with many others in this country, have zealously prosecuted it.

How shall we intelligently express the vast distances at which the stars are removed from us? Of course we can use miles, and pile up the millions upon millions by adding on ciphers, but that fails to give much notion of the star's distance. Let us try with Alpha Centauri: its parallax of 0".75 means that it is 275,000 times farther from the sun than the earth is. Multiplying this out, we get 25 trillion miles, that is, 25 millions of million miles—an inconceivable number, and an unthinkable distance.

Suppose the entire solar system to shrink so that the orbit of Neptune, sixty times 93 million miles in diameter, would be a circle the size of the dot over this letter i. On the same scale the sun itself, although nearly a million miles in diameter, could not be seen with the most powerful microscope in existence; and on the same scale also we should have to have a circle ten feet in diameter, if the solar system were imagined at its center and Alpha Centauri in its circumference.

So astronomers do not often use the mile as a yardstick of stellar distance, any more than we state the distance from London to San Francisco in feet or inches. By convention of astronomers, the average distance between the centers of sun and earth, or 93 million miles, is the accepted unit of measure in the solar system. So the adopted unit of stellar distance is the distance traveled by a wave of light in a year's time: and this unit is technically called the light-year. This unit of distance, or stellar yardstick, as we may call it, is nearly 6 millions of million miles in length. Alpha Centauri, then, is four and one-third light-years distant, and 61 Cygni seven and one-fifth light-years away.

For convenience in their calculations most astronomers now use a longer unit called the parsec, first suggested by Turner. Its length is equal to the distance of a star whose parallax is one second of arc; that is, one parsec is equal to about three and a quarter light-years. Or the light-year is equal to 0.31 parsec. Also the parsec is equal to 206,000 astronomical units, or about 19 millions of million miles.

We have, then four distinct methods of stating the distance of a star: Sirius, for example, has a parallax of 0″.38 or its distance is two and two-thirds parsecs, or eight and a half light-years, or 50 millions of million miles. It is the angle of parallax which is always found first by actual measurement and from this the three other estimates of distance are calculated.

So difficult and delicate is the determination of a stellar distance that only a few hundred parallaxes have been ascertained in the past century. The distance of the same star has been many times measured by different astronomers, with much seeming duplication of effort. Comprehensive campaigns for determining star parallaxes in large numbers have been undertaken in a few instances, particularly at the suggestion of Kapteyn, the eminent astronomer of Groningen, Holland. His catalogue of star parallaxes is the most complete and accurate yet published, and is the standard in all statistical investigations of the stars.

That we find relatively large parallaxes for some of the fainter stars, and almost no measurable parallax for some of the very bright stars is one of the riddles of the stellar universe. We may instance Arcturus, in the northern hemisphere and Canopus in the southern; the latter almost as bright as Sirius. Dr. Elkin and the late Sir David Gill determined exhaustively the parallax of Canopus, and found it very minute, only 0″.03, making its distance in excess of a hundred light-years. The stupendous brilliancy of this star is apparent if we remember that the intensity of its light must vary inversely as the square of the distance; so that if Canopus were to be brought as near us as even 61 Cygni is, it would be a hundredfold brighter than Sirius, the brightest of all the stars of the firmament.

In researches upon the distribution of the more distant stars, the method of measuring parallaxes of individual stars fails completely, and the secular parallax, or parallactic motion of the stars is employed instead. By parallactic motion is meant the apparent displacement in consequence of the solar motion which is now known with great accuracy, and amounts to 19.5 kilometers per second. Even in a single year, then, the sun's motion is twice the diameter of the earth's orbit, so that in a hundred or more years, a much longer base line is available than in the usual type of observations for stellar parallax. If we ascertain the parallactic motion of a group of stars, then we can find their average distance. It is found, for example, that the mean parallax of stars of the sixth magnitude is 0".014. Also the mean distances of stars thrown into classes according to their spectral type have been investigated by Boss, Kapteyn, Campbell and others. The complete intermingling of the two great star streams has been proved, too, by using the magnitude of the proper motions to measure the average distances of both streams. These come out essentially the same, so that the streaming cannot be due to mere chance relation in the line of sight.

Most unexpected and highly important is the discovery that the peculiar behavior of certain lines in the spectrum leads to a fixed relation between a star's spectrum and its absolute magnitude, which provides a new and very effective method of ascertaining stellar distances. By absolute magnitudes are meant the magnitudes the stars would appear to have if they were all at the same standard distance from the earth.

Very satisfactory estimates of the distance of exceedingly remote objects have been made within recent years by this indirect method, which is especially applicable to spiral nebulæ and globular clusters. The absolute magnitude of a star is inferred from the relative intensities of certain lines in its spectrum, so that the observed apparent magnitude at once enables us to calculate the distance of the star. Adams and Joy have recently determined the luminosities and parallaxes of 500 stars by this spectroscopic method. Of these stars 360 have had their parallaxes previously measured; and the average difference between the spectroscopic and the trigonometric values of the parallax is only the very small angle 0".0037, a highly satisfactory verification.

An indirect method, but a very simple one, and of the greatest value because it provides the key to stellar distances with the least possible calculation, and we can ascertain also the distances of whole classes of stars too remote to be ascertained in any other way at present known.

The problem of spectroscopic determinations of luminosity and parallax has been investigated at Mount Wilson with great thoroughness from all sides, the separate investigations checking each other. A definitive scale for the spectroscopic determination of absolute magnitudes has now been established, and the parallaxes and absolute magnitudes have already been derived for about 1,800 stars.

CHAPTER XLIX

THE NEAREST STARS

Of especial interest are the few stars that we know are the nearest to us, and the following table includes all those whose parallax is 0".20 or greater. There are nineteen in all and nearly half of them are binary systems. The radial motions given are relative to the sun. The transverse velocities are formed by using the measured parallaxes to transform proper motions into linear measures. They are given by Eddington in his "Stellar Movements":

Star's Name	Magnitude	Parallax in Seconds of Arc	Proper Motion in Seconds of Arc	Linear Velocity Km. per sec.	Radial Velocity Km. per sec.	Spectral Type	Luminosity (Sun=1)	Star Stream
Groombridge 34	8.2	0.28	2.85	48	..	Ma	0.010	I
Eta Cassiop	3.6	0.20	1.25	30	+10	F8	1.4	I
Tau Ceti	3.6	0.33	1.93	28	-16	K	0.50	II
Epsilon Erid	3.3	0.31	1.00	15	+16	K	0.79	II
CZ 5h 243	8.3	0.32	8.70	129	+242	G-K	0.007	II
Sirius	-1.6	0.38	1.32	16	-7	A	48.0	II
Procyon	0.5	0.32	1.25	19	-3	F5	9.7	I ?
Lal. 21185	7.6	0.40	4.77	57	..	Ma	0.009	II
Lal. 21258	8.9	0.20	4.46	106	..	Ma	0.011	I
OA (N) 11677	9.2	0.20	3.03	72	..	..	0.008	I

Alpha Centauri	0.3	0.76	3.66	23	-22	G,K5	2.0 0.6	I
OA (N) 17415	9.3	0.27	1.31	23	..	F	0.004	II
Pos. Med. 2164	8.8	0.29	2.28	37	..	K	0.006	I
Sigma Draco	4.8	0.20	1.84	43	+25	K	0.5	II
Alpha Aquilæ	0.9	0.24	0.65	13	-33	A5	12.3	I
61 Cygni	5.6	0.31	5.25	80	-39	K5	0.10	I
Epsilon Indi	4.7	0.28	4.67	79	-62	K5	0.25	I
Krüger 60	9.2	0.26	0.92	17	..	..	0.005	II
Lacaille 9352	7.4	0.29	7.02	115	+12	Ma	0.019	I

These stars are distant less than five parsecs (about 16 light-years) from the sun, so they make up the closest fringe of the stellar universe immediately surrounding our system. The large number of binary systems is quite remarkable. Why some stars are single and others double is not yet known. By the spectroscopic method the proportion is not so large; Campbell finding that about one quarter of 1,600 stars examined are spectroscopic binaries, and Frost two-fifths to a half. The exceptional number of large velocities is very remarkable; the average transverse motion of the nineteen stars is fifty kilometers per second, whereas thirty is about what would have been expected.

As to star streams to which these nearest stars belong, eleven are in Stream I and eight in Stream II, in close accord with the ratio 3:2 given by the 6,000 stars of Boss's catalogue. "We are not able," says Eddington, "to detect any significant difference between the luminosities, spectra, or speeds of the stars constituting the two streams. The thorough interpenetration of the two star streams is well illustrated, since we find even in this small volume of space that members of both streams are mingled together in just about the average proportion."

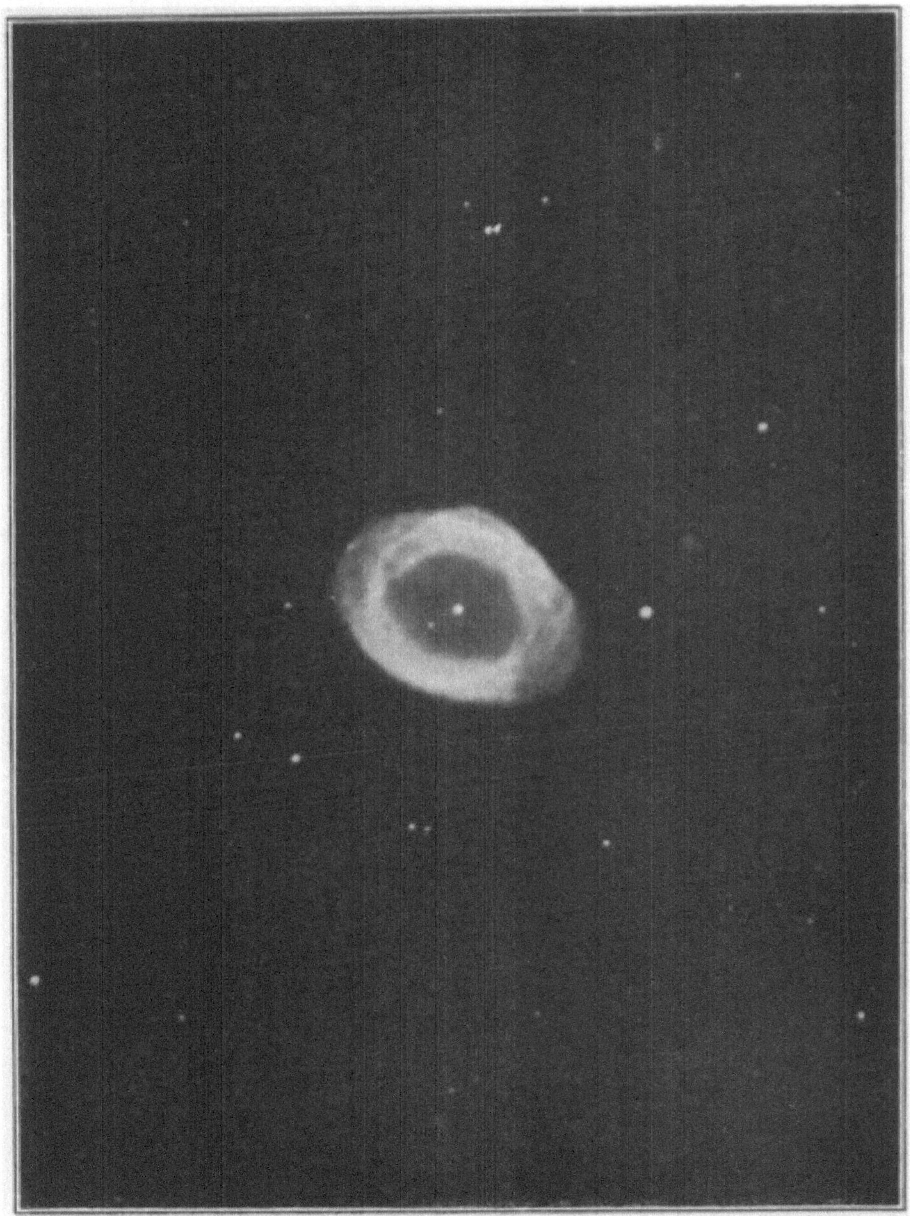

The Ring Nebula in Lyra. This is the best example of the annular and elliptic nebulæ, which are not very abundant.

(Photo, Mt. Wilson Solar Observatory.)

The Dumb-bell Nebula of Vulpecula. To take the photograph required an exposure of five hours.

(Photo, Mt. Wilson Solar Observatory.)

CHAPTER L

ACTUAL DIMENSIONS OF THE STARS

We have seen that the distances of the stars from the solar system are immense beyond conception, and millions upon millions of them are probably forever beyond our power of ascertaining by direct measurement what their distance really is. After we had found the sun's distance and measured the angle filled by his disk, it was easy to calculate his actual size. This direct method, however, fails when we try to apply it to the stars, because their distances are so vast that no star's disk fills an angle of any appreciable size; and even if we try to get a disk with the highest magnifying powers of a great telescope our efforts end only in failure. There is, indeed, no instrumentally appreciable angle to measure.

How then shall we ascertain the actual dimensions of the vast spheres which we know the stars actually are, as they exist in the remotest regions of space? Clearly by indirect methods only, and it must be said that astronomers have as yet no general method that yields very satisfactory results for stellar dimensions. The actual magnitude of the variable system of Algol, Beta Persei, is among the best known of all the stars, because the spectroscope measures the rate of approach and recession of Algol when its invisible satellite is in opposite parts of the orbit; the law of gravitation gives the mass of the star and the size of its orbit, and so the length of the eclipse gives the actual size of the dark, eclipsing body. This figures out to be practically the same size as that of our sun, while Algol's own diameter is rather larger, exceeding a million miles.

If we try to estimate sizes of stars by their brightness merely, we are soon astray. Differences of brightness are due to difference of dimensions, of course, or of light-giving area; but differences of distance also affect the brightness, inversely as the squares of the distances, while differences of temperature and constitution affect, in very marked degree, the intrinsic brilliance of the light-emitting surface of the star. There are big stars and little stars, stars relatively near to us and stars exceedingly remote, and stars highly incandescent as well as others feebly glowing.

We have already shown how the angular diameters subtended by many of the stars have been estimated, through the relation of surface brightness and spectral type. Antares and Betelgeuse appear to be the most inviting for investigation, because their estimated angular diameters are about one-twentieth of a second of arc. This is the way in which their direct measurement is being attempted.

As early as 1890, Michelson of Chicago suggested the application of interference methods to the accurate measurement of very small angles, such as the diameters of the minor planets, and the satellites of Jupiter and Saturn, as well as the arc distance between the components of double stars. Two portions of the object glass are used, as far apart as possible on the same diameter, and the interference fringes produced at the focus of the objective are then the subject of observation. These fringes form a series of equidistant interference bands, and are most distinct when the light comes from a source subtending an infinitesimal angle. If the object presents an appreciable angle, the visibility is less and may even become zero.

Michelson tested this method on the satellites of Jupiter at the Lick Observatory in 1891, and showed its accuracy and practicability. Nevertheless, the method has not been taken up by astronomers, until very recently at the Mount Wilson Observatory, where Anderson has applied it to the measurement of close double stars. It is found that, contrary to general expectation, the method gives excellent results, even if the "seeing" is not the best—2 on a scale of 10, for instance.

To simplify the manipulation of the interferometer, a small plate with two apertures in it is placed in the converging beam of light coming from the telescope objective or mirror. The interference fringes formed in the focal plane are then viewed with an eyepiece of very high power, many thousand diameters. The resolving power of the interferometer is found to be somewhat more than double that of a telescope of the same aperture. By applying the interferometer method to Capella, arc distances of much less than one-twentieth of a second of arc were measured. More recently the method has been applied to the great star Betelgeuse in Orion, whose angular diameter was found to be 0".46, corresponding to an actual diameter of 260,000,000 miles, if the star's parallax is as small as it appears to be.

CHAPTER LI

THE VARIABLE STARS

Spectacular as they are to the layman, novæ, or temporary stars, are to the astronomers simply a class among many thousands of stars which they call variables, or variable stars. There are a few objects classified as irregular variables, one of which is very remarkable. We refer to Eta Argus, an erratic variable in the southern constellation Argo and surrounded by a well-known nebula. There is a pretty complete record of this star. Halley in 1677 when observing at Saint Helena recorded Eta Argus as of the fourth magnitude. During the 18th century, it fluctuated between the fourth magnitude and the second. Early in the 19th it rapidly waxed in brightness, fluctuating between the first and second magnitudes from 1822 to 1836. But two years later its light tripled, rivaling all the fixed stars except Canopus and Sirius. In 1843 it was even brighter for a few months, but since then it has declined fairly steadily, reaching a minimum at magnitude seven and a half in 1886, with a slight increase in brightness more recently. A period of half a century has been suggested, but it is very doubtful if Eta Argus has any regular period of variation.

Another very interesting class of variables is known as the Omicron Ceti type. Nearly all the time they are very faint, but quite suddenly they brighten through several magnitudes, and then fade away, more or less slowly, to their normal condition of faintness. But the extraordinary thing is that most of these variables go through their fluctuations in regular periods: from six months to two years in length. The type star, Omicron Ceti, or Mira, is the oldest known variable, having been discovered by Fabricius in 1596. Most of the time it is a relatively faint star of the 12th magnitude; but once in rather less than a year its brightness runs up to the fourth, third and sometimes even the second magnitude, where it remains for a week or ten days, and afterward it recedes more slowly to its usual faintness, the entire rise and decline in brightness usually requiring about 100 days. The spectrum of Omicron Ceti contains many very bright lines, and a large proportion of the variable stars are of this type.

Another class of variables is designated as the Beta Lyræ type. Their periods are quite regular, but there are two or more maxima and minima of light in each period, as if the variation were caused by superposed relations in some way. Their spectra show a complexity of helium and hydrogen bands. No wholly satisfactory explanation has yet been offered. Probably they are double stars revolving in very small orbits compared with their dimensions, their plane of motion passing nearly through the earth.

But the most interesting of all the variables are those of the Algol type, their light curves being just the reverse of the Omicron Ceti type; that is, they are at their maximum brightness most of the time, and then suffer a partial eclipse for a relatively brief interval. Algol goes through its variations so frequently that its period is very accurately known; it is 2d. 20h. 48m. 55.4s. For most of this period Algol is an easy second magnitude star; then in about four and a half hours it loses nearly five-sixths of its light, receding to the fourth magnitude. Here at minimum it remains for fifteen or twenty minutes, and then in the next three and a half hours it regains its full normal brilliancy of the second magnitude. During these fluctuations the star's spectrum undergoes no marked changes. The spectra of all the Algol variables are of the first or Sirian type.

To explain the variation of the Algol type of variables is easy: a dark, eclipsing body, somewhat smaller than the primary is supposed to be traveling round it in an orbit lying nearly edgewise to our line of sight. The gravitation of this dark companion displaces Algol itself alternately toward and from the earth, because the two bodies revolve round their common center of gravity. With the spectroscope this alternate motion of Algol, now advancing and now receding at the rate of 26 miles per second, has been demonstrated; and the period of this motion synchronizes exactly with the period of the star's variability.

Russell and Shapley have made extended studies of the eclipsing binaries, and developed the formulæ by which the investigations of their orbits are conducted. Heretofore, visual binaries and spectroscopic binaries afforded the only means of deriving data regarding double systems, but it is now possible to obtain from the orbits of eclipsing variables fully as much information relating to binary systems in general and their bearing on stellar evolution. After an orbit has been determined from the photometric data of the light curve, the addition of spectroscopic data often permits the calculation of the masses, dimensions and densities in terms of the sun. Shapley's original investigation included the orbits of ninety eclipsing variables, and with the aid of hypothetical parallaxes, he computed the approximate position of each system in space. The relation to the Milky Way is interesting, the condensation into the Galactic plane being very marked; only thirteen of the ninety systems being found at Galactic latitudes exceeding 30 degrees.

If we can suppose the variable stars covered with vast areas of spots, perhaps similar to the spots on the sun, and then combine the variation of these spot areas with rotation of the star on its axis, there is a possibility of explanation of many of the observed phenomena, especially where the range of variation is small. But for the Omicron Ceti type, no better explanation offers than that afforded by Sir Norman Lockyer's collision theory. First he assumes that these stars are not condensed bodies, but still in the condition of meteoric swarms, and the revolution of lesser swarms around larger aggregations, in elliptic orbits of greater or less eccentricity, must produce vast multitudes of collisions; and these collisions, taking place at pretty regular periods, produce the variable maximum light by raising hosts of meteoric particles to a state of incandescence simultaneously.

The catalogues of variable stars now contain many thousands of these objects. They are often designated by the letters R, S, T, and so on, followed by the genitive form of the name of the constellation wherein they are found. Most of the recently found variables have a range of less than one magnitude. They are so distributed as to be most numerous in a zone inclined about 18 degrees to the celestial equator, and split in two near where the cleft in the Galaxy is located. Nearly all the temporary stars are in this duplex region. Bailey of Harvard a quarter century ago began the investigation of variables in close star clusters, where they are very abundant, with marked changes of magnitude within only a few hours.

Many amateur astronomers afford very great assistance to the professional investigator of variable stars by their cooperation in observing these interesting bodies, in particular the American Association of Observers of Variable Stars, organized and directed by William Tyler Olcott.

For a high degree of accuracy in determining stellar magnitudes the photo-electric cell is unsurpassed. Stebbins of Urbana has been very successful in its application and he discovered the secondary minimum of Algol with the selenium cell. His most recent work was done with a potassium cell with walls of fused quartz, perfected after many trial attempts. The stars he has recently investigated are Lambda Tauri, and Pi Five Orionis. Combining results with those reached by the spectroscope, the masses of the two component stars of the former are 2.5 and 1.0 that of the sun, and the radii are 4.8 and 3.6 times the sun's.

Russell of Princeton thinks it probable that similar causes are at work in all these variables. In the case of the typical Novæ there is evidence that when the outburst takes place a shell of incandescent gas is actually ejected by the star at a very high velocity. What may be the forces that cause such an explosion can only be guessed. Repeated outbursts have not, in the case of T Pyxidis, destroyed the star, because it has gone through this process three times in the past thirty years. Russell inclines to regard it as a standard process occurring somewhere in the stellar universe probably as often as once a year.

Novæ, then, cannot be due to collisions between two stars, for even if we suppose the stars to be a thousand millions in number, no two should collide except at average intervals of many million years. The idea is gaining ground that the stars are vast storehouses of energy which they are gradually transforming into heat and radiating into space. "Under ordinary circumstances, it is probable that the rate of generation of heat is automatically regulated to balance the loss by radiation. But it is quite conceivable that some sudden disturbance in the substance of the star, near the surface, might cause an abrupt liberation of a great amount of energy, sufficient to heat the surface excessively, and drive the hot material off into infinite space, in much the form of a shell of gas, as seems to have been observed in the case of Nova Aquilæ.... With the rapid advance of our knowledge of the properties of the stars on one hand, and of the very nuclei of atoms on the other, we may, perhaps before many years have passed, find ourselves nearer a solution of the problem."

The Cepheid variables increase very rapidly in brightness from their least light to their maximum, and then fade out much more slowly, with certain irregularities or roughnesses of their light-curves when declining. Their spectral lines also shift in period with their variations of light. In the case of these variables, whose regular fluctuation of light cannot be due to eclipse, and is as a rule embraced within a few days, there is a fluctuation in color also between maximum and minimum, as if there were a periodic change in the star's physical condition. Eddington and Shapley advocate the theory of a mechanical pulsation of the star as most plausible. Knowledge of the internal conditions of the stars make it possible to predict the period of pulsation within narrow limits; and for Delta Cephei this theoretical period is between four and ten days. Its observed period is five and one-third days, and corresponding agreement is found in all the Cepheids so far tested.

Shapley of Mount Wilson finds that the Cepheid variables with periods exceeding a day in length all lie close to the Galactic lane. So greatly have the studies of these objects progressed that, as before remarked, when we know the star's period, we can get its absolute magnitude, and from this the star's distance. On all sides of the sun, the distances of the Cepheids range up to 4,000 parsecs. So they indicate the existence of a Galactic system far greater in extent than any previously dealt with.

CHAPTER LII
THE NOVÆ, OR NEW STARS

New stars, or temporary stars, we have already mentioned in connection with variables. They are, next to comets, the most dramatic objects in the heavens. They may be variable stars which, in a brief period, increase enormously in brightness, and then slowly wane and disappear entirely, or remain of a very faint stellar magnitude.

In the ancient historical records are found accounts of several such stars. For instance, in the Chinese annals there is an allusion to such a stellar outburst in the constellation of Scorpio, B. C. 134. This was observed also by Hipparchus and, no doubt, it was the immediate incentive which led to his construction of the first known catalogue of stars, so that similar happenings might be detected in the future. In November, 1572, Tycho Brahe observed the most famous of all new stars, which blazed out in the constellation Cassiopeia. In something over a year it had completely disappeared.

In 1604-1605 a new star of equal brightness was seen in Ophiuchus by Kepler; it also faded out to invisibility in 1606. Kepler and Tycho printed very complete records of these remarkable objects. The eighteenth century passed without any new stars being seen or recorded. There was one of the fifth magnitude in 1848, and another of the seventh magnitude in 1860; and in May, 1866, a star of the second magnitude suddenly made its appearance in Corona Borealis; and one of the third magnitude in Cygnus in November, 1876. The latter was fully observed by Schmidt of Athens and became a faint telescopic star within a few weeks. It is now of the fifteenth magnitude.

In 1885 astronomers were surprised to find suddenly a new star of the sixth magnitude very close to the brightest part of the great nebula in Andromeda; it ran its course in about six months, fading with many fluctuations in brightness, and no star is now visible in its position even with the telescope. Stars of this class are known to astronomers as Novæ, usually with the genitive of the constellation name, as Nova Andromedæ.

In 1891-1892 Nova Aurigæ made its spectacular appearance and yielded a distinctly double and complex spectrum for more than a month. Many pairs of lines indicated a community of origin as to substance, and accurate measurement showed a large displacement with a relative velocity of more than 500 miles per second. For each bright hydrogen line displaced toward the red there was a dark companion line or band about equally displaced toward the violet much as if the weird light of Nova Aurigæ originated in a solid globe moving swiftly away from us and plunging into an irregular nebulous mass as swiftly approaching us. Parallax observations of Nova Aurigæ made it immensely remote, perhaps within the Galaxy, and it still exists as a faint nebulous star.

In February, 1901, in the constellation Perseus appeared the most brilliant nova of recent years. It was first discovered by Dr. Anderson, an amateur of Glasgow, and at maximum on February 23 it outshone Capella. There were many unusual fluctuations in its waning brightness. Its spectrum closely resembled that of Nova Aurigæ, with calcium, helium, and hydrogen lines. In August, 1901, an enveloping nebula was discovered, and a month later certain wisps of this nebulosity appeared to have moved bodily, at a speed seventy-fold greater than ever previously observed in the stellar universe.

According to Sir Norman Lockyer's meteoritic hypothesis, a vast nebulous region was invaded, not by one but by many meteor swarms, under conditions such that the effects of collision varied greatly in intensity. The most violent of these collisions gave birth to Nova Persei itself, and the least violent occurred subsequently in other parts of the disturbed nebula, perhaps immeasurably removed. This explanation would avoid the necessity of supposing actual motion of matter through space at velocities heretofore unobserved and inconceivably high. A recent photograph of Nova Persei, by Ritchey, reveals a nebulous ring of regular structure surrounding the star. The great power of the 60-inch has made it possible to photograph even the spectra of many of the novæ of years ago which are now very faint. After the lapse of years the characteristic lines of the nebular spectrum generally vanish, as if the star had passed out of the nebula—a plunge into which is generally thought to be the cause of the great and sudden outburst of light.

Many novæ have recently been found in the spiral nebulæ, especially in the great nebula of Andromeda.

CHAPTER LIII
THE DOUBLE STARS

Examining individual stars of the heavens more in detail, thousands of them are found to be double; not the stars that appear double to the naked eye, as Theta Tauri, Mizar, Epsilon Lyræ, and others; but pairs of stars much closer together, and requiring the power of the telescope to divide or separate them. Only a very few seconds apart they are or, in many cases, only the merest fraction of a second of arc. Some of them, called binaries, are found to be revolving around a common center, sometimes in only a few years, sometimes in stately periods of hundreds of years. Many such binary systems are now known, and the number is constantly increasing. Castor is one, Gamma Virginis another, Sirius also is one of these binaries, and a most interesting one, having a period of revolution of about 52 years.

Aitken, of the Lick Observatory, in his work on binary stars, directs special attention to the correlation between the elements of known binary orbits and the star's spectral type, and presents a statistical study of the distribution of 54,000 visual double stars, of which the spectra of 3919 are known. That the masses of binary systems average about twice that of the sun's mass has long been known, and this fact can be employed with confidence in estimates of the probable parallax of these systems. Aitken applies the test to fourteen visual systems for which the necessary data are available, and deduces for them a mean mass of 1.76 times that of the sun. For the spectroscopic binaries the masses are much greater.

Triple, quadruple and multiple stars are less frequent; but many exceedingly interesting objects of this class exist. Epsilon Lyræ is one, a double-double, or four stars as seen with slender telescopic power, and six or seven stars with larger instruments. Sigma Orionis and 12 Lyncis, also Theta Cancri and Mu Bootis are good examples of triple stars.

CHAPTER LIV

THE STAR CLUSTERS

From multiple stars the transition is natural to star clusters although the gap between these types of stellar objects is very broad. The familiar group of the winter sky known as the Pleiades is a loose cluster, showing relatively very few stars even in telescopes or on photographic plates. The "Beehive," or cluster known as Praesepe in Cancer, and a double group in the sword-handle of Perseus, both just visible to the naked eye, are excellent examples of star clusters of the average type. When the moon is absent, they are easily recognized without a telescope as little patches of nebulous light; but every increase of optical power adds to their magnificence.

Then we come in regular succession to the truly marvelous globular clusters, that for instance in Hercules. Messier 13, a recent photograph of which, taken by Ritchey with the 60-inch reflector on Mount Wilson, reveals an aggregation of more than 50,000 stars. But the finest specimens are in the southern hemisphere. Sir John Herschel spent much time investigating them nearly a century ago at the Cape of Good Hope. His description of the cluster in the constellation of Centaurus is as follows: "The noble globular cluster Omega Centauri is beyond all comparison the richest and largest object of the kind in the heavens. The stars are literally innumerable, and as their total light when received by the naked eye affects it hardly more than a star of the fifth or fourth to fifth magnitude, the minuteness of each star may be imagined."

Others of these clusters are so remote that the separate stars are not distinguishable, especially at the center, and their distances are entirely beyond our present powers of direct measurement, although methods of estimating them are in process of development. If gravitation is regnant among the uncounted components of stellar clusters, as doubtless it is, these stars must be in rapid motion, although our photographs of measurements have been made too recently for us to detect even the slightest motion in any of the component stars of a cluster. The only variations are changes of apparent magnitude, of a type first detected in a large number of stars in Omega Centauri, by Bailey of Harvard, who by comparison of photographs of the globular clusters was the first to find variable stars quite numerous in these objects. Their unexplained variations of magnitude take place with great rapidity, often within a few hours.

There are about a hundred of these globular clusters, and the radial velocities of ten of them have been measured by Slipher and found to range from a recession of 410 to an approach of 225 kilometers per second. These excessive velocities are comparable with those found for the spiral nebulæ. Shapley has estimated the distances of many of these bodies, which contain a large number of variable stars of the Cepheid type. By assuming their absolute magnitudes equal to those of similar Cepheids at known distances, he finds their distance represented by the inconceivably minute parallax of 0″.00012, corresponding to 30,000 light-years. This research also places the globular clusters far outside and independent of our Galactic system of stars. The distribution of the globular clusters has also been investigated, and these interesting objects are found almost exclusively in but one hemisphere of the sky. Its center lies in the rich star clouds of Scorpio and Sagittarius. Success in finding the distances of these objects has made it possible to form a general idea of their distribution in three-dimensional space.

The numerous variable stars in any one cluster are remarkable for their uniformity. Accepting variables of this type as a constant standard of absolute brightness, and assuming that the differences of average magnitude of the variables in different clusters are entirely due to differences of distance, the relative distances of many clusters were ascertained with considerable accuracy. Then it was found that the average absolute magnitude of the twenty-five brightest stars in a cluster is also a uniform standard, or about 1.3 magnitudes brighter than the mean magnitude of the variables. This new standard was employed in ascertaining the distances of other clusters not containing many variables.

Shapley further shows that the linear dimensions of the clusters are nearly uniform, and the proper relative positions in space are charted for sixty-nine of these objects. We can determine the scale of the charts, if we know the absolute brightness of our primary standard—the variable stars; and this is deduced from a knowledge of the distances of variables of the same type in our immediate stellar system.

The most striking of all the globular clusters, Omega Centauri, comes out the nearest; nevertheless it is distant 6.5 kiloparsecs. A kiloparsec is a thousand parsecs, and is the equivalent of 3,256 light-years. At the inconceivable distance of sixty-seven kiloparsecs, or more than 200,000 light-years, is the most remote of the globular clusters, known to astronomers as N.G.C. 7006, from its number in the catalogue which records its position in the sky, the New General Catalogue of nebulæ by Dreyer of Armagh.

The clusters are widely scattered, and their center of diffusion is about twenty kiloparsecs on the Galactic plane toward the region of Scorpio-Sagittarius. Marked symmetry with reference to this plane makes it evident that the entire system of globular clusters is associated with the Galaxy itself. But to conceive of this it is necessary to extend our ideas of the actual dimensions of the Galactic system. Almost on the circumference of the great system of globular clusters our local stellar system is found, and it contains probably all the naked-eye stars, with millions of fainter ones. Its size seems almost diminutive, only about one kiloparsec in diameter. The relative location of our local stellar system shows why the globular clusters appear to be crowded into one hemisphere only.

Shapley suggests that globular clusters can exist only in empty space, and that when they enter the regions of space tenanted by stars, they dissolve into the well-known loose clusters and the star clouds of the Milky Way. Strangely the radial velocities of the clusters already observed show that most of them are traveling toward this region, and that some will enter the stellar regions within a period of the order of a hundred million years.

The actual dimensions of globular clusters are not easy to determine, because the outer stars are much scattered. To a typical cluster, Messier 3, Shapley assigns a diameter of 150 parsecs, which makes it comparable with the size of the stellar cluster to which the sun belongs. Also on certain likely assumptions, he finds that the diameter of the great cluster in Hercules, the finest one in our northern sky, is about 350 parsecs, and its distance no less than 30,000 parsecs; in other words, the staggering distance that light would require 9,750,000 years to travel over. While these distances can never be verified by direct measurement, it lends great weight to the three methods of indirect measurement, or estimation, (1) from the diameter of the image of the clusters, (2) from the mean magnitude of the twenty-five brightest stars, and (3) from the mean magnitude of the short period variables, that they are in excellent agreement.

CHAPTER LV
MOVING CLUSTERS

Recent researches on the proper motions of stars have brought to light many groups of stars whose individual members have equal and parallel velocities. Eddington calls these moving clusters. The component stars are not exceptionally near to each other, and it often happens that other stars not belonging to the group are actually interspersed among them. They may be likened to double stars which are permanent neighbors, with some orbital motion, though exceedingly slow.

The connection is rather one of origin; occurring in the same region of space, perhaps, from a single nebula. They set out with the same motion, and have "shared all the accidents of the journey together." Their equality of motion is intact because any possible deflections by the gravitative pull of the stellar system is the same for both. Mutual attraction may tend to keep the stars together, but their community of motion persists chiefly because no forces tend to interfere with it. In this way physically connected pairs may be separated by very great distances.

So with the moving clusters: their component stars may be widely separate on the celestial sphere, but equality of their motions affords a clue to their association in groups. The Hyades, a loose cluster in Taurus, is a group of thirty-nine stars, within an area of about 15 degrees square, which has been pretty fully investigated, especially by the late Professor Lewis Boss; and no doubt many fainter stars in the same region will ultimately be found to belong to the same group.

If we draw arrows on a chart representing the amount and direction of the proper motions of these stars, these arrows must all converge toward a point. This shows that their motions are parallel in space. It is a relatively compact group, and the close convergence shows that their individual velocities must agree within a small fraction of a kilometer per second. Radial velocity measures of six of the component stars are in very satisfactory accord, giving 45.6 kilometers per second for the entire group.

We can get the transverse velocity, and therefrom the distances of the stars, which are among the best known in the heavens, because the proper motions are very accurately known. The mean parallax of the group by this indirect method comes out 0".025, agreeing almost exactly with the direct determination by photography, 0".023, by Kapteyn, De Sitter, and others.

Eddington concludes that this Taurus group is a globular cluster with a slight central condensation. Its entire diameter is about ten parsecs, and its known motion enables us to trace its past and future history. It was nearest the sun 800,000 years ago, when it was at about half its present distance. Boss calculated that in 65 million years, if the present motion is maintained, this group will have receded so far as to appear like an ordinary globular cluster 20′ in diameter, its stars ranging from the ninth to the twelfth apparent magnitude. We may infer that the motion will likely continue undisturbed, because there are interspersed among the group many stars not belonging to it, and these have neither scattered its members nor sensibly interfered with the parallelism of their motion.

Another moving cluster, the similarity of proper motion of whose component stars was first pointed out by Proctor, is known as the Ursa Major system, which embraces primarily Beta, Gamma, Delta, Epsilon, and Zeta Ursæ Majoris, or five of the seven stars that mark the familiar Dipper. But as many as eight other stars widely scattered are thought to belong to the same system, including Sirius and Alpha Coronæ Borealis. The absolute motion amounts to 28.8 kilometers per second, and is approximately parallel to the Galaxy. Turner has made a model of the cluster, which has the form of a flat disk.

Among stars of the Orion type of spectrum are several examples of moving clusters. The Pleiades together with many fainter stars form another moving cluster; as also do the brighter stars of Orion, together with the faint cloudlike extensions of the great nebula in Orion, whose radial velocity agrees with that of the stars in the constellation. Still another very remarkable moving cluster is in Perseus, first detected by Eddington, and embracing eighteen stars, the brightest of which is Alpha Persei.

The further discovery of moving clusters is most important in the future development of stellar astronomy, because with their aid we can find out the relative distribution, luminosity, and distance of very remote stars. So far the stars found associated in groups are of early types of spectrum; but the Taurus cluster embraces several members equally advanced in evolution with the sun, and in the more scattered system of Ursæ Major there are three stars of Type F.

"Some of these systems," Eddington concludes, "would thus appear to have existed for a time comparable with the lifetime of an average star. They are wandering through a part of space in which are scattered stars not belonging to their system—interlopers penetrating right among the cluster stars. Nevertheless, the equality of motion has not been seriously disturbed. It is scarcely possible to avoid the conclusion that the chance attractions of stars passing in the vicinity have no appreciable effect on stellar motions; and that if the motions change in course of time (as it appears they must do) this change is due, not to the passage of individual stars, but to the central attraction of the whole stellar universe, which is sensibly constant over the volume of space occupied by a moving cluster."

CHAPTER LVI

THE TWO STAR STREAMS

Consider the ships on the Atlantic voyaging between Europe and America: at any one time there may be a hundred or more, all bound either east or west, some moving in interpenetrating groups, individuals frequently passing each other, but rarely or never colliding. We might say, there are two great streams of ships, one moving east and the other west.

Now in place of each ship, imagine a hundred ships, and magnify their distances from each other to the vast distances that the stars are from each other, and all in motion in two great streams as before. This will convey some idea of the relatively recent discovery, called by astronomers "star-streaming."

Early in this century the investigation of moving clusters began to reveal the fact that the motions of the stars were not at random throughout the universe, and about 1904 Kapteyn was the first to show that the stellar motions considered in great groups are very far from being haphazard, but that the stars tend to travel in two great streams, or favored directions. This was ascertained by analyzing the proper motions of stars in the sky, many thousands of them, and correcting all for the effect which the known motion of the sun would have upon them. The corrected motion, or part that is left over, is known as the star's own motion, or *motus peculiaris*.

This important investigation was very greatly facilitated by the general catalogue of 6,188 stars well distributed over the entire sky, the work of the late Professor Boss. It was published by the Carnegie Institution of Washington, and includes all stars down to the sixth magnitude. Boss was very critical in the matter of stellar positions and proper motions and his work is the most accurate at present available. Excluding stars of the Orion type and the known members of moving clusters, Kapteyn's investigation was based on 5,322 stars, which he divided into seventeen regions of the sky, each northern region having an antipodal one in the southern hemisphere.

Mathematical analysis of these regions showed them all in substantial agreement, with one exception, and enabled Kapteyn to draw the conclusion that the stars of one stream, called Drift I, move with a speed of thirty-two kilometers per second, while those of the other, Drift II, travel with a speed of eighteen kilometers per second. Their directions are not, like those of east and west bound ships, 180 degrees from each other, but are inclined at an angle of 100 degrees. Drift I embraces about three-fifths of the stars, and Drift II the remaining two-fifths. Quite as remarkable as the drifts themselves is the fact that the relative motion of the two is very closely parallel to the plane of the Milky Way.

This epochal research has very great significance in all investigations of stellar motions, and it has been verified in various ways, particularly by the Astronomer Royal, Sir Frank Dyson, who limited the stars under consideration to 1,924 in number, but all having very large proper motions. In this way the two streams are even more characteristically marked. But radial velocity determinations afford the ultimate and most satisfactory test, and Campbell has this investigation in hand, classifying the stars in their streaming according to the type.

Type A stars are so far found to be confirmatory. Turning to the question of physical differences between the stars of the two streams, Eddington inquires into the average magnitude of the stars in both drifts, and their spectral type. Also whether they are distributed at the same distance from the sun, and in the same proportion in all parts of the sky. His conclusion is that there is no important difference in the magnitudes of the stars constituting the two drifts. Regarding their spectra, stars of early and late types are found in both streams, with a somewhat higher proportion of late types among the stars of Drift II than those of Drift I. Campbell and Moore of the Lick Observatory have investigated seventy-three planetary nebulæ which exhibit the phenomena of star-streaming, and have motions which are characteristic of the stars.

Dealing with the very important question whether the two streams are actually intermingled in space, Eddington finds them nearly at the same mean distance and thoroughly intermingled, and there is no possible hypothesis of Drifts I and II passing one behind the other in the same line of sight. A third drift, to which all the Orion stars belong, is under investigation, together with comprehensive analysis of the drifts according to the spectral type of all the stars included.

The farther research on star-streaming is pushed, the more it becomes evident that a third stream, called Drift O, is necessary, especially to include B-type stars. The farther we recede from the sun, the more this drift is in evidence. At the average distances of B-type stars, the observed motions are almost completely represented by Drift O alone. Halm of Cape Town concludes from recent investigations that the double-drift phenomena (Drifts I and II) is of a distinctly *local* character, and concerns chiefly the stars in the vicinity of the solar system; while stars at the greatest distances from the sun belong preeminently to Drift O.

The 60-inch reflector on Mount Wilson gathers sufficient light so that the spectra of very faint stars can be photographed, and a discussion of velocities derived in this manner has shown that Kapteyn's two star streams extend into space much farther than it was possible to trace them with the nearer stars. Star-streaming, then, may be a phenomenon of the widest significance in reference to the entire universe.

As to the fundamental causes for the two opposite and nearly equal star streams, it is early perhaps to even theorize upon the subject. Eddington, however, finds a possible explanation in the spiral nebulæ, which are so numerous as to indicate the certainty of an almost universal law compelling matter to flow in these forms. Why it does so, we cannot be said to know; but obviously matter is either flowing into the nucleus from the branches of the spiral, or it is flowing out from the nucleus into the branches. Which of the two directions does not matter, because in either case there would be currents of matter in opposite directions at the points where the arms merge in the central aggregation. The currents continue through the center, because the stars do not interfere with one another's paths. As Eddington concludes: "There then we have an explanation of the prevalence of motions to and fro in a particular straight line; it is the line from which the spiral branches start out. The two star streams and the double-branched spirals arise from the same cause."

CHAPTER LVII

THE GALAXY OR MILKY WAY

Grandest of all the problems that have occupied the mind of man is the distribution of the stars throughout space. To the earliest astronomers who knew nothing about the distances of the stars, it was not much of a problem because they thought all the fixed stars were attached to a revolving sphere, and therefore all at essentially the same distance; a very moderate distance, too. Even Kepler held the idea that the distances of individual stars from each other are much less than their distances from our sun.

Thomas Wright, of Durham, England, seems to have been the first to suggest the modern theory of the structure of the stellar universe, about the middle of the eighteenth century. His idea was taken up by Kant who elaborated it more fully. It is founded on the Galaxy, the basal plane of stellar distribution, just as the ecliptic is the fundamental circle of reference in the solar system.

What is the Galaxy or Milky Way?

Here is a great poet's view of the most poetic object in all nature:

> A broad and ample road, whose dust is gold,
> And pavement stars, as stars to thee appear
> Seen in the Galaxy, that Milky Way
> Which nightly as a circling zone thou seest
> Powder'd with stars.

Milton, P. L. vii, 580.

Were the earth transparent as crystal, so that we could see downward through it and outward in all directions to the celestial sphere, the Galaxy or Milky Way would appear as a belt or zone of cloudlike luminosity extending all the way round the heavens. As the horizon cuts the celestial sphere in two, we see at anyone time only one-half of the Milky Way, spanning the dome of the sky as a cloudlike arch.

As the general plane of the Galaxy makes a large angle with our equator, the Milky Way is continually changing its angle with the horizon, so that it rises at different elevations. One-half of the Milky Way will always be below our horizon, and a small region of it lies so near the south pole of the heavens that it can never be seen from medium northern latitudes.

Galileo was the first to explain the fundamental mystery of this belt, when he turned his telescope upon it and found that it was not a continuous sheet of faint light, as it seemed to be, but was made up of countless numbers of stars, individually too faint to be visible to the naked eye, but whose vast number, taken in the aggregate, gave the well-known effect which we see in the sky. In some regions, as Perseus, the stars are more numerous than in others, and they are gathered in close clusters. The larger the telescope we employ, the greater the number of stars that are seen as we approach the Galaxy on either side; and the farther we recede from the Galaxy and approach either of its poles fewer and fewer stars are found. Indeed, if all the stars visible in a 12-inch telescope could be conceived as blotted out, nearly all the stars that are left would be found in the Galaxy itself.

The naked eye readily notes the variations in breadth and brightness of the galactic zone. Nearly a third of it, from Scorpio to Cygnus, is split into two divisions nearly parallel. In many regions its light is interrupted, especially in Centaurus, where a dark starless region exists, known as the "coal sack." Sir John Herschel, who followed up the stellar researches of his father, Sir William, in great detail, places the north pole of the Galactic plane in declination 37 degrees N., and right ascension 12 h. 47 m. This makes the plane of the Milky Way lie at an angle of about 60 degrees with the ecliptic, which it intersects not far from the solstices.

Now Kant, in view of the two great facts about the Galaxy known in his time, (1) that it wholly encircles the heavens, and (2) that it is composed of countless stars too faint to be individually visible to the naked eye, drew the safe conclusions that the system of the stars must extend much farther in the direction of the Milky Way than in other directions.

This theory of Kant was next investigated from an observational standpoint by Sir William Herschel, the ultimate goal of whose researches was always a knowledge of the construction of the heavens. The present conclusion is that we may regard the stellar bodies of the sidereal universe as scattered, without much regard to uniformity, throughout a vast space having in general the shape of a thick watch, its thickness being perhaps one-tenth its diameter. On both sides of this disk of stars, and clustered about the poles of the sidereal system are the regions occupied by vast numbers of nebulæ. The entire visible universe, then, would be spheroidal in general shape. The plane of the Milky Way passes through the middle of this aggregation of stars and nebulæ, and the solar system is near the center of the Milky Way. Throughout the watch-form space the stars are clustered irregularly, in varied and sometimes fantastic forms, but without approach to order or system. If we except some of the star groups and star clusters and consider only the naked-eye stars, we find them scattered with fair approach to uniformity.

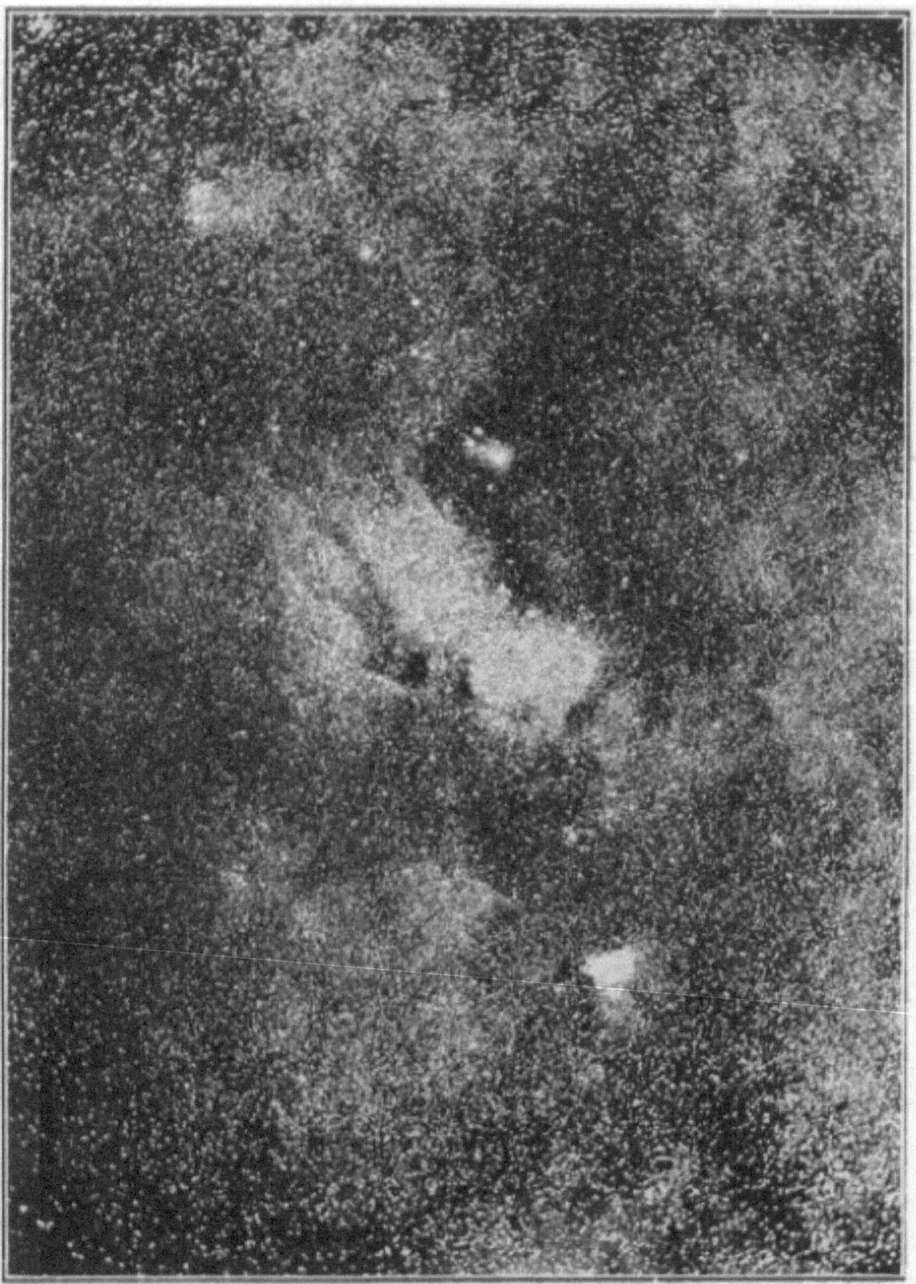

Star Clouds and Black Holes in Sagittarius. The dark rifts and lanes resemble those in the nearby Milky Way.

(Photo, Yerkes Observatory.)

The Great Nebula of Andromeda, Largest (Apparently) of all the Spiral Nebulæ. This nebula can be seen very faintly with the naked eye, but no telescope has yet resolved it into separate stars.

(Photo, Yerkes Observatory.)

The watch-shaped disk is not to be understood as representing the actual form of the stellar system, but only in general the limits within which it is for the most part contained.

A vigorous attack on the problem of the evolution and structure of the stellar universe as a whole is now being conducted by cooperation of many observatories in both hemispheres. It is known as the Kapteyn "Plan of Selected Areas," embracing 206 regions which are distributed regularly over the entire sky. Besides this a special plan includes forty-six additional regions, either very rich or extremely poor in stars, or to which other interest attaches.

Of all investigators Kapteyn has gone into the question of our precise location in the Milky Way most thoroughly, concluding that the solar system lies, not at the center in the exact plane, but somewhat to the north of the Galaxy. Discussing the Sirian stars he finds that if stars of equal brightness are compared, the Sirians average nearly three times more distance from the sun than those of the solar type. So, probably, the Sirians far exceed the Solars in intrinsic brightness. Farther, Kapteyn concludes that the Galaxy has no connection with our solar system, and is composed of a vast encircling annulus or ring of stars, far exceeding in number the stars of the great central solar cluster, and everywhere exceedingly remote from these stars, as well as differing from them in physical type and constitution. So it would be mainly the mere element of distance that makes them appear so faint and crowded thickly together into that gauzy girdle which we call the Galaxy.

The Milky Way reveals irregularities of stellar density and star clustering on a large scale, with deep rifts between great clouds of stars. Modern photographs, particularly those of Barnard in Sagittarius, make this very apparent. Within the Milky Way, nearly in its plane and almost central, is what Eddington terms the inner stellar system, near the center of which is the sun. Surrounding it and near its plane are the masses of star clouds which make up the Milky Way. Whether these star clouds are isolated from the inner system or continuous with it, is not yet ascertained.

The vast masses of the Milky Way stars are very faint, and we know nothing yet as to their proper motions, their radial motions, or their spectra. Probably a few stars as bright as the sixth magnitude are actually located in the midst of the Milky Way clusters, the fainter ninth magnitude stars certainly begin the Milky Way proper, while the stars of the twelfth or thirteenth magnitude carry us into the very depths of the Galaxy.

It is now pretty generally believed that many of the dark regions of the Milky Way are due not to actual absence of stars so much as to the absorption of light by intervening tracts of nebulous matter on the hither side of the Galactic aggregations and, probably in fact, within the oblate inner stellar system itself. Easton has made many hundred counts of stars in galactic regions of Cygnus and Aquila where the range of intensity of the light is very marked; in fact, the star density of the bright patches of the Galaxy is so far in excess of the density adjacent and just outside the Milky Way, that the conclusion is inevitable that this excess is due to the star clouds.

Of the distance of the Milky Way we have very little knowledge. It is certainly not less than 1,000 parsecs, and more likely 5,000 parsecs, a distance over which light would travel in about 16,000 years. Quite certainly all parts of the Galaxy are not at the same distance, and probably there are branches in some regions that lie behind one another. While the general regions of the nebulæ are remote from the Galactic plane, the large irregular nebulæ, as the Trifid, the Keyhole, and the Omega nebulæ, are found chiefly in the Milky Way.

In addition to the irregular nebulæ many types of stellar objects appear to be strongly condensed toward the Milky Way, but this may be due to the inner stellar system, rather than a real relation to Galactic formations. Quite different are the Magellanic clouds, which contain many gaseous nebulæ and are unique objects of the sky, having no resemblance to the true spiral nebulæ which, as a rule, avoid the Galactic regions. Worthy of note also is the theory of Easton that the Milky Way has itself the form of a double-branched spiral, which explains the visible features quite well, but is incapable of either disproof or verification. The central nucleus he locates in the rich Galactic region of Cygnus, with the sun well outside the nucleus itself. By combining the available photographs of the Galaxy, he has produced a chart which indicates in a general way how the stellar aggregations might all be arrayed so as to give the effect of the Galaxy as we see it.

Shapley, at Mount Wilson, has studied the structure of the Galactic system, in which he has been aided by Mrs. Shapley. An interesting part of this work relates to the distribution of the spiral nebulæ, and to certain properties of their systematic recessional motion, suggesting that the entire Galactic system may be rapidly moving through space. Apparently the spiral nebulæ are not distant stellar organizations or "island universes," but truly nebular structures of vast volume which in general are actively repelled from stellar systems. A tentative cosmogonic hypothesis has been formulated to account for the motions, distribution, and observed structure of clusters and spiral nebulæ.

An additional great problem of the Galaxy is a purely dynamical one. Doubtless it is in some sort of equilibrium, according to Eddington, that is to say, the individual stars do not oscillate to and fro across the stellar system in a period of 300 million years, but remain concentrated in clusters as at present. Poincaré has considered the entire Milky Way as in stately rotation, and on the assumption that the total mass of the inner stellar system is 1,000,000,000 times the sun's mass, and that the distance of the Milky Way is 2,000 parsecs, the angular velocity for equilibrium comes out 0".5 per century. That is to say, a complete revolution would take place in about 250 million years.

CHAPTER LVIII
STAR CLOUDS AND NEBULÆ

From star clusters to nebulæ, only a century ago, the transition was thought to be easy and immediate. Accuracy in determining the distances of stars was just beginning to be reached, the clusters were obviously of all degrees of closeness following to the verge of irresolvability, and it was but natural to jump to the conclusion that the mystery of the nebulæ consisted in nothing but their vaster distance than that of clusters, and it was believed that all nebulæ would prove resolvable into stars whenever telescopes of sufficiently great power could be constructed.

But the development of the spectroscope soon showed the error of this hypothesis, by revealing bright lines in the nebular spectra showing that many nebulæ emit light that comes from glowing incandescent gas, not from an infinitude of small stars.

In pre-telescope days nothing was known about the nebulæ. The great nebula in Andromeda, and possibly the great nebula in Orion, are alone visible to the naked eye, but as thus seen they are the merest wisps of light, the same as the larger clusters are. Galileo, Huygens and other early users of the telescope made observations of nebulæ, but long-focus telescopes were not well adapted to this work. Simon Mayer has left us the first drawing of a nebula, the Orion nebula as he saw it in 1612. The vast light-gathering power of the reflectors built by Sir William Herschel first afforded glimpses of the structure of the nebulæ, and if his drawings are critically compared with modern ones, no case of motion with reference to the stars or of change in the filaments of the nebulæ themselves has been satisfactorily made out.

Only very recently has the distance of a nebula been determined, and the few that have been measured seem to indicate that the nebulæ are at distances comparable with the stars. Of all celestial objects the nebulæ fill the greatest angles, so that we are forced to conclude, with regard to the actual size of the greater nebulæ as they exist in space, that they far surpass all other objects in bulk.

Photography invaded the realm of the nebulæ in 1880, when Dr. Henry Draper secured the first photograph of the nebula of Orion. Theoretically photography ought to help greatly in the study of the nebulæ, and enable us in the lapse of centuries to ascertain the exact nature of the changes which must be going on. The differences of photographic processes, of plates, of exposure and development produce in the finished photograph vastly greater differences than any actual changes that might be going on, so that we must rely rather on optical drawings made with the telescope, or on drawings made by expert artists from photographs with many lengths of exposure on the same object.

The great work on nebulæ and star clusters recently concluded by Bigourdan of the Paris Observatory and published in five volumes received the award of the gold medal of the Royal Astronomical Society. While D'Arrest measured about 2,000 nebulæ, and Sir John Herschel about double that number in both hemispheres, Bigourdan has measured about 7,000. His work forms an invaluable lexicon of information concerning the nebulæ.

Classification of the nebulæ is not very satisfactory, if made by their shapes alone. There are perhaps fifteen thousand nebulæ in all that have been catalogued, described, and photographed. Dreyer's new general catalogue (N.G.C.) is the best and most useful. Many of the nebulæ, especially the large ones, can only be classified as irregular nebulæ. The Orion nebula is the principal one of this class, revealing an enormous amount of complicated detail, with exceptional brilliancy of many regions and filaments. An extraordinary multiple star, Theta Orionis, occupies a very prominent position in the nebula, and photographs by Pickering have brought to light curved filaments, very faint and optically invisible, in the outlying regions which give the Orion nebula in part a spiral character. But the delicate optical wisps of this nebula are well seen, even in very small telescopes. Its spectrum yields hydrogen, helium and nitrogen. The Orion nebula is receding from the earth about eleven miles in every second. Keeler and Campbell have shown that nearly every line of the nebular spectrum is a counterpart of a prominent dark line in the spectrum of the brighter stars of the constellation of Orion. A recent investigator of the distribution of luminosity in the great nebula of Orion finds that radiations from nebulium are confined chiefly to the Huygenian region of the nebula and its immediate neighborhood.

Photography has revealed another extraordinary nebula or group of nebulæ surrounding the stars in the Pleiades, which the deft manipulation of Barnard has brought to light. All the stars and the nebula are so interrelated that they are obviously bound together physically, as the common proper motion of the stars also appears to show. Also in the constellation Cygnus, Barnard has discovered very extensive nebulosities of a delicate filmy cloudlike nature which are wholly invisible with telescopes, but very obvious on highly sensitive plates with long exposures.

Another class of these objects are the annular and elliptic nebulæ which are not very abundant. The southern constellation Grus, the crane, contains a fine one, but by far the best example is in the constellation Lyra. It is a nearly perfect ring, elliptic in figure, exceedingly faint in small telescopes; but large instruments reveal many stars within the annulus, one near the center which, although very faint to the eye, is always an easy object on the photographic plate, because it is rich in blue and violet rays. The parallax of the ring nebula in Lyra comes out only one-sixth of that of the planetary nebulæ, and the least greatest diameters of this huge continuous ring are 250 and 330 times the orbit of Neptune.

Planetary nebulæ and nebulous stars are yet another class of nebulæ, for the most part faint and small, resembling in some measure a planetary disk or a star with nebulous outline. Practically all are gaseous in composition, and have large radial velocities. Probably they are located within our own stellar system. The parallaxes of several of them have been measured by Van Maanen: one of the very small angle 0".023, which enables us to calculate the diameter of this faint but interesting object as equal to nineteen times the orbit of Neptune.

CHAPTER LIX
THE SPIRAL NEBULÆ

Last and most important of all are the spiral nebulæ. The finest example is in the constellation Canes Venatici, and its spiral configuration was first noted by Lord Rosse, an epoch-making discovery. The convolutions of its spiral are filled with numerous starlike condensations, themselves engulfed in nebulosity. Photography possesses a vast advantage over the eye in revealing the marvelous character of this object, an inconceivably vast celestial whirlpool. Naturally the central regions of the whorl would revolve most swiftly, but no comparison of drawings and photographs, separated by intervals of many years, has yet revealed even a trace of any such motion.

The number of large spiral nebulæ is not very great; the largest of all is the great nebula of Andromeda, whose length stretches over an arc of seven times the breadth of the moon, and its width about half as great. This nebula is a naked-eye object near Eta Andromedæ, and it is often mistaken for a comet. Optically it was always a puzzle, but photographs by Roberts of England first revealed the true spiral, with ringlike formations partially distinct, and knots of condensing nebulosity as of companion stars in the making. While its spectrum shows the nongaseous constitution of this nebula, no telescope has yet resolved it into component stars.

Systematic search for spiral nebulæ by Keeler, and later continued by Perrine, at the Lick Observatory, with the 36-inch Crossley reflector, disclosed the existence of vast numbers of these objects, in fact many hundreds of thousands by estimation; so that, next to the stars, the spiral nebulæ are by far the most abundant of all objects in the sky. They present every phase according to the angle of their plane with the line of sight, and the convolutions of the open ones are very perfectly marked. Many are filled with stars in all degrees of condensation, and the appearance is strongly as if stars are here caught in every step of the process of making.

The vast multitude of the spiral nebulæ indicates clearly their importance in the theory of the cosmogony, or science of the development of the material universe. Curtis of the Lick Observatory has lately extended the estimated number of these objects to 700,000. He has also photographed with the Crossley reflector many nebulæ with lanes or dark streaks crossing them longitudinally through or near the center. These remarkable streaks appear as if due to opaque matter between us and the luminous matter of the nebula beyond. Perhaps a dark ring of absorptive or occulting matter encircles the nebula in nearly the same plane with the luminous whorls. Duncan has employed the 60-inch Mount Wilson reflector in photographing bright nebulæ and star clusters in the very interesting regions of Sagittarius. One of these shows unmistakable dark rifts or lanes in all parts of the nebula, resembling the dark regions of the neighboring Milky Way.

Pease of Mount Wilson has recently employed the 60-inch and the 100-inch reflectors of the Mount Wilson Observatory to good advantage in photographing several hundred of the fainter nebulæ. Many of these are spirals, and others present very intricate and irregular forms. A search was made for additional spirals among the smaller nebulæ along the Galaxy, but without success. Several of the supposedly variable nebulæ are found to be unchanging. Many nights in each month when the moon is absent are devoted to a systematic survey of the smaller nebulæ and their spectra by photography. The visible spiral figure of all these objects is a double-branched curve, its two arms joining on the nucleus in opposing points, and coiling round in the same geometrical direction. The spiral nebulæ, as to their distribution, are remote from the Galaxy, and the north Galactic polar region contains a greater aggregation than the south. The distances of the spiral nebulæ are exceedingly great. They lie far beyond the planetary and irregular gaseous nebulæ, like that of Orion, which are closely related to the stars forming part of our own system. Possibly the spiral nebulæ are exterior or separate "island universes." If so, they must be inconceivably vast in size, and would develop, not into solar systems, but into stellar clusters. The enormous radial velocities of the spiral nebulæ, averaging 300 to 400 kilometers per second, or twenty-fold that of the stars, tend to sustain the view that they may be "island universes," each comparable in extent with the universe of stars to which our sun belongs.

Recent spectroscopic observations of the nebulæ applying the principle of Doppler have revealed high velocities of rotation. Slipher of the Lowell Observatory made the first discovery of this sort and Van Maanen of Mount Wilson has detected in the great Ursa Major spiral, No. 101 in Messier's catalogue, a speed of rotation at five minutes of arc from the center that would correspond to a complete period in 85,000 years. As was to be expected, the nebula does not rotate as a rigid body, but the nearer the center the greater the angular velocity, and Van Maanen finds evidence of motion along the arms and away from the center.

These great velocities appear to belong to the spiral nebulæ as a class, and not to other nebulæ. Thirteen nebulæ investigated by Keeler are as a whole almost at rest relatively to our system, as are the large irregular objects in Orion, and the Trifid nebula. This would seem to indicate that the spiral nebulæ form systems outside our own and independent of it.

Quite different from the spirals in their distribution through space are the planetary nebulæ. The spirals follow the early general law of nebulæ arrangement, that is, they are concentrated toward the poles of the Galaxy; but the planetary nebulæ, on the other hand, are very few near the poles and show a marked frequency toward the Galactic plane. Campbell and Moore have found spectroscopic evidence of internal rotatory motion in a large proportion of the planetary nebulæ.

The distribution of the nebulæ throughout space, like that of the stars, is still under critical investigation, but the location of vast numbers of the more compact nebulæ on the celestial sphere is very extraordinary. The Milky Way appears to be the determining plane in both cases; the nearer we approach it the more numerous the stars become, whereas this is the general region of fewest nebulæ and they increase in number outward in both directions from the Galaxy, and toward both poles of the Galactic circle. Obviously this relation, or contra-relation of stars and nebulæ on such a vast scale is not accidental, and it also must be duly accounted for in the true theory of the cosmogony. The nebulæ which are found principally in and near the Milky Way are the large irregular nebulæ, and vast nebulous backgrounds, like those photographed by Barnard in Scorpio, Taurus and elsewhere, as well as the Keyhole, Omega, and Trifid nebulæ. Allied to these backgrounds are doubtless some of the dark Galactic spaces, radiating little or no intrinsic light, and absorbing the light of the fainter stars beyond them. A peculiar veiled or tinted appearance has been remarked in some cases visually, and examination of the photographs strongly confirms the existence of absorbing nebulosity.

The spiral nebulæ are so abundant, and so much attention is now being given to them, both by observers and mathematicians, that their precise relation to the stellar systems must soon be known; that is, whether they are comparatively small objects belonging to the stellar system, or independent systems on the borders of the stellar system, or as seems more likely, vast and exceedingly remote galaxies comparable with that of the Milky Way itself. Our knowledge of the motions of the spirals, both radial and angular, is increasing rapidly, and must soon permit accurate general conclusions to be drawn.

CHAPTER LX
COSMOGONY

Down to the middle of the last century and later, it was commonly believed that in the beginning the cosmos came into being by divine fiat substantially as it is. Previously the earth had been "without form and void," as in the Scripture. Had it not been for the growth and gradual acceptance of the doctrine of evolution, and its reactionary effect upon human thought, it is conceivable that the early view might have persisted to the present day; but now it is universally held that everything in the heavens above and the earth beneath is subject more or less to secular change, and is the result of an orderly development throughout indefinite past ages, a progressive evolution which will continue through indefinite aeons of the future.

In the writings of the Greek philosophers, and down through the Middle Ages we find the idea of an original "chaos" prevailing, with no indication whatever of the modern view of the process by which the cosmos came to be what they saw it and as it is to-day. If we go still farther back, there is no glimmer of any ideas that will bear investigation by scientific method, however interesting they may be as purely philosophical conceptions. Many ancient philosophers, among them Anaxagoras, Democritus, and Anaximenes, regarded the earth as the product of diffused matter in a state of the original chaos having fallen together haphazard, and they even presumed to predict its future career and ultimate destiny.

In Anaximander and Anaximenes alone do we find any conception of possible progress; their thought was that as the world had taken time to become what it is, so in time it would pass, and as the entire universe had undergone alternate renewal and destruction in the past, that would be its history in the future. Aristotle, Ptolemy, and others appear to have held the curious notion that although everything terrestrial is evanescent, nevertheless the cosmos beyond the orbit of the moon is imperishable and eternal.

By tracing the history of the intellectual development of Europe we may find why it was that scientific speculation on the cosmogony was delayed until the 18th century, and then undertaken quite independently by three philosophers in three different countries. Swedenborg, the theologian, set down in due form many of the principles that underlie the modern nebular hypothesis. Thomas Wright of Durham whose early theory of the arrangement of stars in the Galaxy we have already mentioned, speculated also on the origin and development of the universe, and his writings were known to Kant, who is now regarded as the author of the modern nebular hypothesis. This presents a definite mechanical explanation of the development and formation of the heavenly bodies, and in particular those composing the solar system.

Kant was illustrious as a metaphysician, but he was a great physicist or natural philosopher as well, and he set down his ideas regarding the cosmogony with precision. Learned in the philosophy of the ancients, he did not follow their speculative conceptions, but merely assumed that all the materials from which the bodies of the solar system have been fashioned were resolved into their original elements at the beginning, and filled all that part of space in which they now move. True, this is pretty near the chaos of the Greeks, but Kant knew of the operation of the Newtonian law of gravitation, which the Greeks did not.

As a natural result of gravitative processes, Kant inferred that the denser portions of the original mass would draw upon themselves the less dense portions, whirling motions would be everywhere set up, and the process would continue until many spherical bodies, each with a gaseous exterior in process of condensation, had taken the place of the original elements which filled space. In this manner Kant would explain the sameness in direction of motion, both orbital and axial, of all the planets and satellites of our system. But many philosophers are of the opinion that Kant's hypothesis would result, not in the formation of such a collection of bodies as the solar system is, but rather in a single central sun formed by common gravitation toward a single center.

From quite another viewpoint the work of the elder Herschel is important here. No one knew the nebulæ from actual observation better than he did; but, while his ideas about their composition were wrong, he nevertheless conceived of them as gradually condensing into stars or clusters of stars. And it was this speculative aspect of the nebulæ, not as a possible means of accounting for the birth and development of the solar system, which constitutes Herschel's chief contribution to the nebular hypothesis. Classifying the nebulæ which he had carefully studied with his great telescopes, it seemed obvious to him that they were actually in all the different stages of condensation, and subsequent research has strongly tended to substantiate the Herschelian view.

Then came Laplace, who took up the great hypothesis where Kant and Herschel had left it, added new and important conceptions in the light of his mature labors as mathematician and astronomer, and put the theory in definitive form, such that it has ever since been known under the name of Laplacian nebular hypothesis. For reasons like those that prevailed with Kant, he began the evolution of the solar system with the sun already formed as the center, but surrounded by a vast incandescent atmosphere that filled all the space which the sun's family of planets now occupy. This entire mass, sun, atmosphere, and all, he conceived to have a stately rotation about its axis. With rotation of the mass and slow reduction of temperature in its outer regions, there would be contraction toward the solar center, and an increase in velocity of rotation until the whole mass had been much reduced in diameter at its poles and proportionately expanded at its equator.

When the centrifugal force of the outer equatorial masses finally became equal to the gravitational forces of the central mass, then these conjoined outer portions would be left behind as a ring, still revolving at the velocity it had acquired when detached. The revolution of the entire inner mass goes on, its velocity accelerating until a similar equilibration of forces is again reached, when a second rotating ring is left behind. Laplace conceived the process as repeated until as many rings had been detached as there are individual planets, all central about the sun, or nearly so.

In all, then, we should have nine gaseous rings; the outer ones preceding the inner in formation, but not all existing as rings at the same time. Radiation from the ring on all sides would lead to rapid contraction of its mass, so that many nuclei of condensation would form, of various sizes, all revolving round the central sun in practically the same period. Laplace conceived the evolution of the ring to proceed still farther till the largest aggregation in it had drawn to itself all the other separate nuclei in the ring.

This, then, was the planet in embryo, in effect a diminutive sun, a secondary incandescent mass endowed with axial rotation in the same direction as the parent nebula. With reduction of temperature by radiation, polar contraction and equatorial expansion go on, and planetary rings are detached from this secondary mass in exactly the same way as from the original sun nebula. And these planetary rings are, in the Laplacian hypothesis, the embryo moons or planetary satellites, all revolving round their several planets in the same direction that the planets revolve about the sun.

In the case of one of the planetary rings, its formation was so nearly homogeneous throughout that no aggregation into a single satellite was possible; all portions of the ring being of equal density, there was no denser region to attract the less dense regions, and in this manner the rings of Saturn were formed, in lieu of condensation into a separate satellite. Similarly in the case of the primal solar ring that was detached next after the Jovian ring; there was such a nice balancing of masses and densities that, instead of a single major planet, we have the well-known asteroidal ring, composed of innumerable discrete minor planets.

This, then, in bare outline, is the Laplacian nebular hypothesis, and it accounted very well for the solar system as known in his day; the fairly regular progression of planetary distances; their orbits round the sun all nearly circular and approximately in a single plane; the planetary and satellite revolutions in orbit all in the same direction; the axial rotations of planets in the same direction as their orbital revolutions; and the plane of orbital revolution of the satellites practically coinciding with the plane of the planet's axial rotation. But the principle of conservation of energy was, of course, unknown to Laplace, nor had the mechanical equivalence of heat with other forms of energy been established in his day.

In 1870, Lane of Washington first demonstrated the remarkable law that a gaseous sphere, in process of losing heat by radiation and contraction because of its own gravity, actually grows hotter instead of cooler, as long as it continues to be gaseous, and not liquid or solid. So there is no need of postulating with Laplace an excessively high temperature of the original nebula. The chief objection to Laplace's hypothesis by modern theorists is that the detachment of rings, though possible, would likely be a rare occurrence; protuberances or lumps on the equatorial exterior of a swiftly revolving mass would be more likely, and it is much easier to see how such masses would ultimately become planets than it is to follow the disruption of a possible ring and the necessary steps of the process by which it would condense into a final planet. The continued progress of research in many departments of astronomy has had important bearing on the nebular hypothesis, and we may rest assured that this hypothesis in somewhat modified form can hardly fail of ultimate acceptance, though not in every essential as its great originator left it.

Lord Rosse's discovery of spiral nebulæ, followed up by Keeler's photographic search for these bodies, revealing their actual existence in the heavens by the hundreds of thousands, has led to another criticism of the Laplacian theory. Could Laplace have known of the existence of these objects in such vast numbers, his hypothesis would no doubt have been suitably modified to account for their formation and development. It is generally considered that the ring of Saturn suggested to Laplace the ring feature in his scheme of origin of planets and satellites; so far as we know, the Saturnian ring is unique, the only object of its kind in the heavens. Whereas, next to the star itself, the spiral nebula is the type object which occurs most frequently. A theory, therefore, which will satisfactorily account for the origin and development of spiral nebulæ must command recognition as of great importance in the cosmogony.

Such a theory has been set forth by Chamberlin and Moulton in their planetesimal hypothesis, according to which the genesis of spiral nebulæ happens when two giant suns approach each other so closely that tide-producing effects take place on a vast scale. These suns need not be luminous; they may perhaps belong to the class of dark or extinguished suns. The evidences of the existence of such in vast numbers throughout the universe is thought to be well established.

Now, on close approach, what happens? There will be huge tides, and the nearer the bodies come to each other, the vaster the scale on which tides will be formed. If the bodies are liquid or gaseous, they will be distorted by the force of gravitation, and the figure of both bodies will become ellipsoidal; and at last under greater stress, the restraining shell of both bodies will burst asunder on opposite sides in streams of matter from the interior. In this manner the arms of the spiral are formed.

As Chamberlin puts it: "If, with these potent forces thus nearly balanced, the sun closely approaches another sun, or body of like magnitude ... the gravity which restrains this enormous elastic power will be reduced along the line of mutual attraction. At the same time the pressure transverse to this line of relief will be increased. Such localized relief and intensified pressure must bring into action corresponding portions of the sun's elastic potency, resulting in protuberances of corresponding mass and high velocity."

Only a fraction of one per cent of the sun's mass ejected in this fashion would be sufficient to generate the entire planetary system. Nuclei or knots in the arms of the spiral gradually grew by accretion, the four interior knots forming Mercury, Venus, the Earth, and Mars. The earth knot was a double one, which developed into the earth-moon system. The absence of a dominating nucleus beyond Mars accounts for the zone of the asteroids remaining in some sense in the original planetesimal condition. The vaster nuclei beyond Mars gradually condensed into Jupiter, Saturn, Uranus, and Neptune; and lesser nuclei related to the larger ones form the systems of moons or satellites.

The orbits of the planetesimals and the planetary and satellite nuclei would be very eccentric, forming a confusion of ellipses with frequently crossing paths. Collisions would occur, and the nuclei would inevitably grow by accretion. Each planet, then, would clear up the planetesimals of its zone; and Moulton shows that this process would give rise to axial revolution of the planet in the same direction as its orbital revolution. The eccentricities would finally disappear, and the entire mass would revolve in a nearly circular orbit.

Rotation twists the streams into the spiral form, and the huge amounts of wreckage from the near-collision are thrown into eddies. The fragments or particles (planetesimals) which have given the name to the theory, begin their motion round their central sun in elliptical paths as required by gravitation. The form of the spiral is preserved by the orbital motion of its particles. There is a gradual gathering together of the planetesimals at points or nodes of intersection, and these become aggregations of matter, nuclei that will perhaps become planets, though more likely other stars. The appulse or near approach is but one of the methods by which the spiral nebulæ may have come into existence. The planetesimal hypothesis would seem to account for the formation of many of these objects as we see them in the sky, though perhaps it is hardly competent to replace entirely the Laplacian hypothesis of the formation of the solar system, which would appear to be a special case by itself.

It will be observed that while the Laplacian hypothesis is concerned in the main with the progressive development of the solar system, and systems of a like order surrounding other stellar centers, whose existence is highly probable, the origin and development of the stellar universe is a vaster problem which can only be undertaken and completed in its broadest bearings when the structure of the stellar universe has been ascertained.

Darwin's important investigations in 1877-1878 on tidal friction may be here related. Before his day acceptance of the ring-theory of development of the moon from the earth had scarcely been questioned; but his recondite mathematical researches on the tidal reaction between a central yielding mass and a body revolving round it brought to light the unsuspected effect of tides raised upon both bodies by their mutual attraction. The type of tides here meant is not the usual rise and fall of the waters of the ocean, but primeval tides in the plastic material of which the earth in its early history was composed. The Newtonian law of gravitation afforded a complete explanation of the rise and fall of the waters of the oceans, but as applied to the motions of planets and satellites by the Lagrangian formulæ, it presupposed that all these bodies are rigid and unyielding. However, mutual tides of phenomenal height in their early plastic substances must have been a necessary consequence of the action of the Newtonian law, and they gradually drew upon the earth's rotational moment of momentum.

In its very early history, before there was any moon to produce tides, the earth rotated much more rapidly, that is, the day was very much shorter than now, probably about five or six hours long. And with the rapid whirling, it was not a Laplacian ring that was detached, but a huge globular mass was separated from the plastic earth's equator. Darwin shows that the gravitative interaction of the two bodies immediately began to raise tides of extraordinary height in both, therefore tending to slow down the rotational periods of both bodies. Action and reaction being equal, the reaction at once began driving the moon away from the earth and thereby lengthening its period of revolution. So small was the mass of the moon and so near was it to the earth, that its relative rotational energy was in time completely used up, and the moon has ever since turned her constant face toward us. Tides of sun and moon in the plastic earth, acting through the ages, slowed down the earth's rotation to its present period, or the length of the day.

Moulton, however, has investigated the tidal theory of the origin of the moon in the light of the planetesimal hypothesis, concluding that the moon never was part of the earth and separated therefrom by too rapid rotation of the earth, but that the distance of the two bodies has always been the same as now. The more massive earth has in its development throughout time robbed the less massive moon in the gradual process of accretion. So the moon has never acquired either an ocean or atmosphere, and this view is acceptable to geologists who have studied the sheer lunar surface, Shaler of Harvard among the first, and laid the foundations for a separate science of selenology.

Tidal friction has also been operant in producing sun-raised tides upon the early plastic substances which composed the planets: more powerfully in the case of planets nearer the sun; less rapidly if the planet's mass is large; also less completely if the planet has solidified earlier on account of its small dimensions. So Darwin would account for the present rotation periods of all the planets: both Mercury and Venus powerfully acted on by the sun on account of their nearness to him, and their rotational energy completely exhausted, so that they now and for all time turn a constant face toward him, as the moon does to the earth; earth and possibly Mars even yet undergoing a very slight lengthening of their day; Jupiter and Saturn, also Uranus and probably Neptune, still exhibiting relatively swift axial rotation, because of their great mass and great original moment of momentum, and also by reason of their vast distances from the central tide-raising body, the sun.

By applying to stellar systems the principles developed by Darwin, See accounted for the fact, to which he was the first to direct attention, that the great eccentricity of the binary orbits is a necessary result of the secular action of tidal friction. The double stars, then, were double nebulæ, originally single, but separated by a process allied to that known as "fission" in protozoans. Indeed, Poincaré proved mathematically that a swiftly revolving nebula, in consequence of contraction, first undergoes distortion into a pear-shaped or hour-glass figure, the two masses ultimately separating entirely; and the observations of the Herschels,

Lord Rosse and others, with the recent photographic plates at the Lick and Mount Wilson observatories, afford immediate confirmation in a multitude of double nebulæ, widely scattered throughout the nebular regions of the heavens.

Jeans of Cambridge, England, among the most recent of mathematical investigators of the cosmogony, balances the advantages and disadvantages of the differing cosmogonic systems as follows, in his "Problems of Cosmogony and Stellar Dynamics": "Some hundreds of millions of years ago all the stars within our Galactic universe formed a single mass of excessively tenuous gas in slow rotation. As imagined by Laplace, this mass contracted owing to loss of energy by radiation, and so increased its angular velocity until it assumed a lenticular shape.... After this, further contraction was a sheer mathematical impossibility and the system had to expand. The mechanism of expansion was provided by matter being thrown off from the sharp edge of the lenticular figure, the lenticular center now forming the nucleus, and the thrown-off matter forming the arms, of a spiral nebula of the normal type. The long filaments of matter which constituted the arms, being gravitationally unstable, first formed into chains of condensation about nuclei, and ultimately formed detached masses of gas. With continued shrinkage, the temperature of these masses increased until they attained to incandescence, and shone as luminous stars. At the same time their velocity of rotation increased until a large proportion of them broke up by fission into binary systems. The majority of the stars broke away from their neighbors and so formed a cluster of irregularly moving stars—our present Galactic universe, in which the flattened shape of the original nebula may still be traced in the concentration about the Galactic plane, while the original motion along the nebular arms still persists in the form of 'star-streaming.' In some cases a pair or small group of stars failed to get clear of one another's gravitational attractions and remain describing orbits about one another as wide binaries or multiple stars. The stars which were formed last, the present B-type stars, have been unusually immune from disturbance by their neighbors, partly because they were born when adjacent stars had almost ceased to interfere with one another, partly because their exceptionally large mass minimized the effect of such interference as may have occurred; consequently they remain moving in the plane in which they were formed, many of them still constituting closely associated groups of stars—the moving star clusters.

"At intervals it must have happened that two stars passed relatively near to one another in their motion through the universe. We conjecture that something like 300 million years ago our sun experienced an encounter of this kind, a large star passing within a distance of about the sun's diameter from its surface. The effect of this, as we have seen, would be the ejection of a stream of gas toward the passing star. At this epoch the sun is supposed to have been dark and cold, its density being so low that its radius was perhaps comparable with the present radius of Neptune's orbit. The ejected stream of matter, becoming still colder by radiation, may have condensed into liquid near its ends and perhaps partially also near its middle. Such a jet of matter would be longitudinally unstable and would condense into detached nuclei which would ultimately form planets."

CHAPTER LXI

COSMOGONY IN TRANSITION

We have seen how Wright in 1750 initiated a theory of evolution, not only of the solar system, but of all the stars and nebulæ as well; how Kant in 1752 by elaborating this theory sought to develop the details of evolution of the solar system on the basis of the Newtonian law, though weakened, as we know, by serious errors in applying physical laws; how Laplace in 1796 put forward his nebular hypothesis of origin and development of the solar system, by contraction from an original gaseous nebula in accord with the Newtonian law; how Sir William Herschel in 1810 saw in all nebulæ merely the stuff that stars are made of; how Lord Rosse in 1845 discovered spiral nebulæ; how Helmholtz in 1854 put forward his contraction theory of maintenance of the solar heat, seemingly reinforcing the Laplacian theory; how Lane in 1870 proved that a contracting gaseous star might rise in temperature; how Roche in 1873 in attempting to modify the Laplacian hypothesis, pointed out the conditions under which a satellite would be broken up by tidal strains; how Darwin in 1879 showed that the theory of tidal evolution of non-rigid bodies might account for the formation of the moon, and binary stars might originate by fission; how Keeler in 1900 discovered the vast numbers of spiral nebulæ; how Chamberlin and Moulton in 1903 put forward the planetesimal hypothesis of formation of the spiral nebulæ, showing also how that hypothesis might account for the evolution of the solar system; and how Jeans in 1916 advocated the median ground in evolution of the arms of the spiral nebulæ, showing that they will break up into nuclei, if sufficiently massive.

In all these theories, truth and error, or lack of complete knowledge, appear to be intermingled in varying proportions. Is it not early yet to say, either that any one of them must be abandoned as totally wrong, or on the other hand that any one of them, or indeed any single hypothesis, can explain all the evolutionary processes of the universe?

Clearly the great problems cannot all be solved by the kinetic theory of gases and the law of gravitation alone. Recent physical researches into sub-atomic energy and the structure and properties of matter, appear to point in the direction where we must next look for more light on such questions as the origin and maintenance of the sun's heat, the complex phenomena of variable stars and the progressive evolution of the myriad bodies of the stellar universe. Because we have actually seen one star turn into a nebula we should not jump to the conclusion that all nebulæ are formed from stars, even if this might seem a direct inference from the high radial velocities of planetary nebulæ.

Quite as obviously many of the spiral nebulæ are in a stage of transition into local universes of stars—even more obvious from the marvelous photographs in our day than the evolution of stars from nebulæ of all types was to Herschel in his day.

The physicist must further investigate such questions as the building up of heavy atomic elements by gravitative condensation of such lighter ones as compose the nebulæ; and laboratory investigation must elucidate further the process of development of energy from atomic disintegration under very high pressures. This leads to a reclassification of the stars on a temperature basis.

Equally important is the inquiry into the mechanism of radiative equilibrium in sun and stars. Not impossibly the process of the earth's upper atmosphere in maintaining a terrestrial equilibrium may afford some clue. What this physical mechanism may be is very incompletely known, but it is now open to further research through recent progress of aeronautics, which will afford the investigator a "ceiling" of 50,000 feet and probably more. Beneath this level, perhaps even below 40,000 feet, lie all the strata, including the inversion layer, where the sun's heat is conserved and an equilibrium maintained.

Even ten years ago, had an astronomer been asked about the physical condition of the interior of the stars, he would have replied that information of this character could only be had on visiting the stars themselves—and perhaps not even then. But at the Cardiff meeting of the British Association in 1920, Eddington, the president of Section A, delivered an address on the internal constitution of the stars. He cites the recent investigations of Russell and others on truly gaseous stars, like Aldebaran, Arcturus, Antares and Canopus, which are in a diffuse state and are the most powerful light-givers, and thus are to be distinguished from the denser stars like our Sun. The term *giants* is applied to the former, and *dwarfs* to the latter, in accord with Russell's theory.

As density increases through contraction, these terms represent the progressive stages, from earlier to later, in a star's history. A red or M-type star begins its history as a giant of comparatively low temperature. Contracting, according to Lane's law, its temperature must rise until its density becomes such that it no longer behaves as a perfect gas. Much depends on the star's mass; but after its maximum temperature is attained, the star, which has shrunk to the proportions of a dwarf, goes on cooling and contracts still further.

Each temperature-level is reached and passed twice, once during the ascending stage and once again in descending—once as a giant, and once as a dwarf. Thus there are vast differences in luminosity: the huge giant, having a far larger surface than the shrunken dwarf, radiates an amount of light correspondingly greater.

The physicist recognizes heat in two forms—the energy of motion of material atoms, and the energy of ether waves. In hot bodies with which we are familiar, the second form is quite insignificant; but in the giant stars, the two forms are present in about equal proportions. The super-heated conditions of the interior of the stars can only be estimated in millions of degrees; and the problem is not one of convection currents, as formerly thought, bringing hot masses to the surface from the highly heated interior, but how can the heat of the interior be barred against leakage and reduced to the relatively small radiation emitted by the stars. "Smaller stars have to manufacture the radiant heat which they emit, living from hand to mouth; the giant stars merely leak radiant heat from their store."

So a radioactive type of equilibrium must be established, rather than a convective one. Laboratory investigations of the very short waves are now in progress, bearing on the transparency of stellar material to the radiation traversing it; and the penetrating power of the star's radiation is much like that of X-rays. The opacity is remarkably high, explaining why the star is so nearly "heat-tight."

Opacity being constant, the total radiation of a giant star depends on its mass only, and is quite independent of its temperature or state of diffuseness. So that the total radiation of a star which is measured roughly by its luminosity, may readily remain constant during the entire 'giant' stage of its history. As Russell originally pointed out, giant stars of every spectral type have nearly the same luminosity. From the range of luminosity of the giant stars, then, we may infer their range of masses: they come out much alike, agreeing well with results obtained by double-star investigation.

These studies of radiation and internal condition of the stars again bring up the question of the original source of that supply of radiant energy continually squandered by all self-luminous bodies. The giant stars are especially prodigal, and radiate at least a hundredfold faster than the sun.

"A star is drawing on some vast reservoir of energy," says Eddington, "by means unknown to us. This reservoir can scarcely be other than the sub-atomic energy which, it is known, exists abundantly in all matter; we sometimes dream that man will one day learn how to release it and use it for his service. The store is well-nigh inexhaustible, if only it could be tapped. There is sufficient in the sun to maintain its output of heat for fifteen billion years."

Lector House believes that a society develops through a two-fold approach of continuous learning and adaptation, which is derived from the study of classic literary works spread across the historic timeline of literature records. Therefore, we aim at reviving, repairing and redeveloping all those inaccessible or damaged but historically as well as culturally important literature across subjects so that the future generations may have an opportunity to study and learn from past works to embark upon a journey of creating a better future.

This book is a result of an effort made by Lector House towards making a contribution to the preservation and repair of original ancient works which might hold historical significance to the approach of continuous learning across subjects.

HAPPY READING & LEARNING!

LECTOR HOUSE LLP
E-MAIL: lectorpublishing@gmail.com

www.ingramcontent.com/pod-product-compliance
Lightning Source LLC
Chambersburg PA
CBHW031452160726
47994CB00005B/2001